Biomimetic Membrane Technology

Donald Martin • Isabelle Vilgrain

Biomimetic Membrane Technology

Biological Engineering and Nanomedicine

Springer

Donald Martin
Laboratoire des Matériaux et du Génie Physique (UMR 5628)
Université Grenoble Alpes,
CNRS-Grenoble INP
Grenoble, France

Isabelle Vilgrain
Unité Mixte de Recherche INSERM-CEA-UGA-CNRS U13 / Institut de Recherches Interdiscipinaires de Grenoble
Laboratoire Biosciences et Bioingenierie pour la Santé-BIOMICS
CEA-Grenoble / Grenoble Cedex, France

ISBN 978-3-031-90593-3 ISBN 978-3-031-90594-0 (eBook)
https://doi.org/10.1007/978-3-031-90594-0

This Springer imprint is published by the registered company Springer Nature Switzerland AG
The registered company address is: Gewerbestrasse 11, 6330 Cham, Switzerland

Preface

The initiating event which ultimately led to publication of this book was that there were a few textbooks that focused on the significance of biomimetic membranes in the diagnostic and therapeutic technologies applied to nanomedicine. Our original objective for this book was to fill that need and to provide a comprehensive but focused volume on the initiation and development of biomimetic membrane technology in the fields of biological engineering and nanomedicine. Writing this book has allowed us the opportunity to consolidate information in fields in which there has been great activity, but published across several disciplines and in diverse primary publications. While this book is not strictly a second edition of our Springer title *Nanobiotechnology of Biomimetic Membranes* (ISBN-10: 0-387-37738-7), this book describes ways to utilize biomimetic membranes in assembling nanostructured systems that have usefulness in nanomedicine.

Biomimetic membrane technology has brought enormous potential value in our efforts to understand biological functions and molecular interactions not only in normal development and cellular physiology but also in diseases. The growing interest of scientists, clinicians, and engineers for nanobiotechnology has opened new fascinating questions to understand the fundamental cores of how biological systems work. Since nanoscaled devices provide measurement opportunities at molecular scales, utilizing the approaches we describe could enhance the investigators to understand how cells respond to the ever-changing environment of the body and how proteins in cell membranes and the intracellular signaling pathways modulate the biochemical and genetic environment inside the cell.

Biomimetic membrane technology opens new pathways to biologically engineer a smart nano-structured system, known as a symbio-bot, which is considered to be in the domain of nanobiotechnology. In this context, the cell membrane and membrane-associated proteins are of particular importance. The present volume is positioned as a response to the research question proposed by Richard Feynman in reply to a suggestion made by his colleague Albert R. Hibbs concerning the feasibility of constructing miniature machines capable of being permanently integrated into the human body to facilitate the functioning of a compromised organ. The question and challenge posed by Richard Feynman was "How do we make such a tiny mechanism? I leave that to you," which can be found in the text published in 1960 from his Caltech lecture titled "There's plenty of room at the bottom."

Biomimetic membrane technology refers to our passion for development of engineered tools that are dedicated to biomedicine, and we hope that the readers of our book will share our enthusiasm for the pursuit of knowledge with regard to the functionality of biological systems. We thank Merry Stuber and her team at Springer Nature for their patience and support during the writing of this book. We hope that this book will stimulate and facilitate the creation of smart nanostructured systems that will benefit the practice of nanomedicine.

Grenoble, France Isabelle Vilgrain
La Tronche, France Donald Martin

Contents

Chapter 1
What Is This Thing Called Biomimetic Membrane Technology

Is there anything more to discover? This is a question that scientists, clinicians, and engineers ask of themselves during their pursuit of the ways to explain the function of natural systems. Just when you think it is safe to turn off the computer, there is yet another publication of a new twist to an explanation for a physical, chemical, or biological function. The new twists are usually found at smaller and smaller observational scales, and hence the appetite for scientists, clinicians, and engineers to apply nanoscale measurements to their investigations. When applied to biomedical problems, this need for nanoscaled measurements is generally considered in the realm of nanobiotechnology, where the function of the cell membrane and membrane-incorporated proteins are important. Such studies have been previously described in the context of the nanobiotechnology of biomimetic membranes [1].

The purpose of this current book is to discover how applications of biomimetic membrane technology can be applied to develop novel medical diagnostic and therapeutic devices. This draws biomimetic membrane technology into one of the central themes that are important to the growth of the field of nanomedicine. Nature is full of complex things that are elegant in the simplicity of their methods of fabrication. These methods usually include the self-assembly of molecules that is triggered by the environmental conditions of the cell, including the characteristics of the cellular fluids (e.g., phase changes, pH, temperature, ion concentration, and redox state). Inspiration from such elegant natural processes provides a path for biomimetic membrane technology to be a basis for novel medical diagnostic and therapeutic devices. The overarching principle that is at the core of this path for biomimetic membrane technology is the way in which nanotechnology can be inspired by nature to self-assemble functional biomimetic membranes that are based on lipid bilayer membranes. That overarching principle of nanobiotechnology is a thread that will be developed and expanded in this book. We will utilize an extensive bibliometric analysis of the state-of-the-art published documents in the fields of nanobiotechnology and nanomedicine to investigate that thread. Then, we will weave this thread into the novel concept of biological engineering and describe how

D. Martin, I. Vilgrain, *Biomimetic Membrane Technology*,
https://doi.org/10.1007/978-3-031-90594-0_1

this concept underpins the perspective of the future directions for biomimetic membrane technology in advancing the fields of nanobiotechnology and nanomedicine.

There is a dichotomy of whether the starting point for advances in nanobiotechnology and nanomedicine is in the realm of fundamental scientists to first seek knowledge in the basic functions of molecules that comprise biomimetic membranes or in the realm of clinicians and engineers who first seek to observe and characterize defined medical problems and needs. In those scenarios, the fundamental scientist will have the subsequent task to find a biomedical application of such fundamental knowledge and understanding. The clinician or engineer will have the subsequent task to find, or indeed reverse-engineer, a targeted understanding of the functions and principles that underlie the defined problem. This question of the preeminence of fundamental or applied research will continue to be the source of debate amongst scientists, clinicians, and engineers.

Discussing the philosophical superiority of either approach is outside the scope of this book. Nonetheless, the pragmatic aspect of whether there is a fundamental or an applied starting point is important for designing and doing research in nanomedicine. We will resolve those fundamental-versus-applied issues by introducing the approach of biological engineering, which results in multidisciplinary-targeted outcomes from research that requires a team with interdisciplinary expertise. We can commence with the premise that the value of nanomedicine is to improve the health of individuals. A related premise is that the value of biomimetic membrane technology provides a symbiotic interface to improve the function of systems and devices that interact with tissues and cells of the body [2]. This pragmatic approach will be highlighted with examples throughout the discussions in this book. Indeed, such a biological engineering approach allows contributions from both a fundamental and an applied approach within the same project. This is because the multidisciplinary outcomes are the important features driving such a biological engineering approach, and not the specific disciplines. This in turn drives the conceptual structure of the knowledge that is required to best be able to biologically engineer a nanobiotechnological solution. Indeed, these nanobiotechnological advances support the advancement of nanomedicine. We will return to the discussion of nanomedicine at the end of this chapter.

Having said that, and with the view in mind to at least find a starting point for advances in biomimetic membrane technology, we can choose to start in the fundamental scientific fields of physics and chemistry. These fields are the traditional bases that underlie nanotechnology. However, the application of fundamental knowledge from those two fields to the area of biomimetic membrane technology is necessarily guided by an inspiration from nature. Many recent scientific publications state such an inspiration either as explicitly being the basis for a hypothesis and design of experimentation or implicitly in the discussions that relate the fundamental results to a biomedical application. For example, this can be seen in the elegant work on nanoscaled vectors for intracellular delivery and internalization [3] that was based on manipulating the chemistry of block copolymers [4]. Whether such an inspiration from natural processes is either explicitly or implicitly considered in such fundamental studies in chemistry or physics, the consequent advances in biomimetic membrane technology eventually drive the processes and engineering of membranes to develop the application of such fundamental results. In that sense, the fields of bioengineering and biology are also important areas to examine in developing an understanding of the advances in biomimetic membrane technology. Indeed, the application of such fundamental results will also depend on how clinicians will engage with and utilize biomimetic membrane technology to improve the health of people.

Biomimetic Membrane Technology and Biological Principles

Indeed, biomimetic membrane technology is influenced by an understanding of biological principles. The technology of biomimetic membranes relies on integrating chemistry, physical sciences, and biology with a strong influence from the way that natural systems are constructed. Natural systems tend to construct themselves through self-directed assembly. One of the major tasks for a modern nanobiotechnologist is to understand, define, and utilize those "rules" to harness self-assembly to produce reproducible bioinspired biomimetic membranes. An example of those tasks can be illustrated in the development of a functional tethered lipid bilayer membrane system. These biomimetic membrane systems are stable and suitable for the study of membrane proteins. This example also illustrates the dichotomy between where the starting point should be—with the fundamental knowledge or the need to provide a practical solution to a need.

The recent nanobiotechnology of biomimetic membranes can be seen as the culmination of a long history of achievements that commenced with the fabrication of model lipid membranes possessing some degree of predictability, which were first reported in the pioneering work of K.R. Blodgett and I. Langmuir, who formed thin monomolecular films supported on solid or liquid surfaces [5–7]. Prior to that pioneering work, the thickness of suspended biomolecular liquid films of soap in air had been investigated and measured [8, 9]. The extension of that work to produce biomolecular liquid films in air led to an understanding of the formation of lipid films in water [10]. Based on those early reports, a major milestone in suspended lipid bilayer research was achieved by the reconstitution of an excitable cell membrane structure in a saline solution, which was in effect the first report of the production of the black lipid membrane [11, 12]. Those early reports paved the way for the first report of liposomes and functional studies of the ionic transport across swollen phospholipids at the Babraham Laboratories in Cambridge [13]. Those major achievements by pioneering scientists, including E.S. Johonnott, J. Folch, P. Mueller, A.D. Bangham, K.R. Blodgett, and I. Langmuir, created a platform for the development of the understanding of transport processes in lipid bilayer membranes. Recent developments, however, have seen the use of model biomimetic membranes as more than an aid to better understand biological function and have now advanced to become a core engineered element in a new generation of biomimetic molecular devices.

The challenge for any biomimetic membrane technology based on lipid membranes is to ensure the stability and durability of the lipid membranes. Although challenging to engineer, good stability and durability are usually achieved using lipid bilayers that are tethered to or supported by a substrate that may be non-porous (e.g., gold, silicon) or porous (e.g., polymer, gel), rather than using suspended lipid bilayers. The first reported example of a stable and durable biomimetic membrane system that was based on a tethered lipid bilayer utilized an ion channel as a biosensor [14]. This has now been developed into a commercially available platform to provide a turnkey tethered lipid bilayer device for the study of membrane proteins and membrane-based biosensors (Surgical Diagnostics Pty Ltd).

The elegance of the Cornell tethered lipid membrane also stimulated a field of fundamental research to search for alternative molecules and approaches to the concept of a tethered lipid bilayer. Although previous authors had published the formation of tethered lipid bilayers with thiopeptide films [15], or incorporated ATPase to demonstrate enzyme function in such lipid bilayer membranes [16], it can be considered that Cornell defined how to utilize the "rules of lipid interactions" to harness self-assembly to produce reproducible bioinspired biomimetic membranes that were robust and stable for long periods of time to function as ion-transporting biosensors. Cornell also was the first to successfully utilize a membrane-spanning ion channel to transport ions to achieve the ion-channel switch biosensor. This can be considered as an example of establishing a "solution-driven" starting point for such a biomimetic membrane technology. Indeed, since the Cornell publication of 1997, there has been a plethora of alternate approaches to the details for a tethered lipid bilayer, which has been championed by Knoll [17] and followed by several groups related to Knoll's initial studies.

Notwithstanding the philosophy of where the starting point should be—fundamental or applied—the core of biomimetic membrane technology remains as understanding and applying the function of lipid bilayer membranes. The reason is that the lipid membrane is fundamental to the organization of biological molecules within the body. In the following chapters, we emphasize the ways that scientists and technologists can measure and utilize the biomimetic functions of lipid bilayer membranes. One of the most important of those functions is the way that nature utilizes the lipid bilayer as a selectively permeable barrier for cells in order to achieve control over flows of electrolytes. Those flows of electrolytes are crucial in generating and transmitting electrical signals by nerve and muscle cells, in generating electrochemical potentials in response to external cellular stimuli by specialized sensing cells, in maintaining volume control in cells, and in secreting fluids such as saliva and tears. The selectivity in controlling flows of electrolytes is achieved by the incorporation of transport proteins, sensing proteins, and transduction proteins into the lipid bilayer. With those systems included in the lipid bilayer, nature has provided the cell with an elegant sensing and transduction system.

Cells in the body are enclosed by a lipid bilayer membrane, and indeed the organelles within the cell are also enclosed by a lipid bilayer membrane. It is the membrane transport proteins in the lipid bilayer, especially ion channels, that provide the fundamental electrochemical signals to maintain cell functions, including the creation of action potentials that allow communication in the nervous system. The transmembrane diffusion of electrolytes through ion channels and other membrane transport proteins provides the energy for osmotically driven fluid balance in the body. This includes the fundamental driving forces for the serous component of fluid secretion from exocrine glands and the control of urine production in the kidneys. Drugs need to cross the lipid bilayer of cells in order to be absorbed, enter the blood or other internal fluid compartments, and then provide function by binding with biological target molecules, many of which are membrane-bound transport proteins and ion channels. Also, implanted medical devices interact with the extracellular environment and are invariably enclosed in some form of membrane.

The success of technologies for human medical applications relies on a good understanding of the interaction and incorporation of macromolecules in membranes and, indeed, the fundamental properties of the membrane itself. There is already a large biological literature on the structure and function of lipid membranes in cells. The purpose of this book is not to reproduce that extensive biological literature, but rather to focus on the ways in which biomimetic membrane technology can play a role in medical diagnosis and therapy. The book develops the concepts of natural membranes and related proteins at the scale of the cell, which are then taken to the scale of a person in order to identify diseases where there is a membrane-based protein at the basis. We discuss the ways in which diagnostic and therapeutic devices can be based on biomimetic membrane technology in order to provide better disease-management solutions for the clinician. The book concludes with chapters that describe the advanced microfabrication techniques and research tools that are important for devices based on biomimetic membrane technology.

The following sections of this chapter synthesize some of the published literature in these fields, with the purpose to place a context of fundamental and applied research into biomimetic membrane systems for the further application of that research into the medical framework for diagnostics and therapies. These following sections of this chapter provide an introduction to the biomimetic membrane technology that will be developed in the remaining chapters of this book.

Biomimetic Membranes and Interfaces

The technological advances that are based on an understanding of self-assembly and lipid bilayer structure are in engineering synthetic, or "artificial," solutions to construct engineered biomimetic membrane systems. This bioinspired engineering is not easy. Up to now a successful approach has been to design systems that are a hybrid of artificial and naturally occurring molecules. This type of system is exemplified by the tethered lipid bilayer membrane that was pioneered in the groups of Cornell and Knoll. The next stage of complexity is an entirely synthetic system that comprises synthetic phospholipids and synthetic nanopores, for example, which could be used to control the transport of electrolytes across synthetic bilayers. However, both the hybrid and the entirely synthetic systems have an ongoing role in medical diagnostic and therapeutic applications. For example, a hybrid system will be very useful for investigating the detailed control of transport proteins that are important for disease, with innovative approaches such as cell-free production of proteins directly into hybrid biomimetic membranes [18]. On the other hand, an entirely synthetic system would be important for long-term stable biosensing systems or for implanted applications.

Such an entirely synthetic biomimetic system would provide an engineered membrane that was effectively immortal. The art of self-healing and self-regeneration is a secret of nature that has not yet been understood in sufficient detail to become a routine engineering design tool for bioinspired systems. Synthetic systems provide a next-best alternative to completely self-regenerating biologically based biomimetic systems, with a trite example being the self-healing capabilities of PVA polymers. Nonetheless, the secrets of self-healing, self-regeneration, growth, and metabolism that exist in living cells remain targets for ongoing research. Scientists and engineers have developed a great depth of understanding the biochemical and biophysical processes of several natural cellular systems, but the underlying causal cellular components that provide the essence of "life" remain elusive. For example, this has been the subject of many philosophical discussions, which are succinctly stated in a recent publication [19]. Indeed, the complexity of some biochemical processes still remains elusive. For example, the reconstruction of artificial systems to mimic the energy-converting capability of photosynthesis is a dream that has been pursued by scientists for more than a century. In 1912, Prof Giacomo Ciamician proposed photochemical systems to replace coal as a source of energy, and he hypothesized … "Is *fossil* [sic] solar energy the only one that may be used in modern life and civilization? That is the question" [20]. Despite more than a century of research efforts, a completely functional biomimetic system for artificial photosynthesis has never been constructed. A recent breakthrough has been made with integrated photobioelectrochemical systems that contain novel enzymatic cathodes and photoanodes [21]. Electronically engineered systems (e.g., photovoltaic) for energy conversion also provide a technological alternative, but there is little doubt that the efficiency of energy conversion by natural photosynthesis far exceeds any of the technological solutions that exist today. A biomimetic system to produce energy by membrane-based artificial photosynthesis would provide a major advance in energy production. The bioinspired engineering of such a system could benefit from a "left-field" approach to utilize synthetic components to self-assemble the biomimetic membrane and perhaps also the enzymes (i.e., proteins) that could be incorporated to drive the energy conversion.

If we continue this theme of "immortalized" synthetic biomimetic membrane systems, it is clear that the building blocks could be synthetic lipids to be used in biomimetic systems that mimic the self-assembly processes of the naturally occurring lipid bilayer membrane. A starting point is the reconstitution of lipoprotein nanostructures. The naturally occurring lipoprotein is a self-assembled nanostructure of apolipoprotein and various amphipathic and hydrophobic lipid species. Lipoproteins are distinguished by their mass density, apolipoprotein type, and protein-to-lipid ratio. For example, high density lipoprotein (HDL) is a class of lipoprotein with the highest density as well as the greatest protein-to-lipid ratio and smallest diameter of approximately 10 nm [22]. Synthetic lipoprotein can be self-assembled by incubating purified apoA-I with small unilamellar vesicles (SUVs) comprising pure phospholipids, which forms the reconstituted HDL (rHDL, also referred to as nanodiscs or nanolipoprotein particles, NLPs). The disk-shaped planar bilayer configuration of the rHDL is self-assembled with two or more amphipathic protein molecules forming a looped-belt structure around rHDL. The diameter of the lipid platform section of these HDL reconstitutions is tunable, since the apolipoprotein looped-belt has the ability to form tightly looped segments that limit the maximum possible length of the lipid-bound conformation. This usually occurs in response to variations in the protein-to-lipid ratio; for example, an increase typically results in a smaller diameter lipoprotein complex. Other approaches to production of biomimetic HDL complexes involve custom-manufactured variants of apolipoproteins that offer more precise control over the final size of the particle [22].

Polypeptides that are able to be custom-made also provide building blocks to self-assemble synthetic biomimetic membranes [23]. A large variety of polypeptide sequences can be manufactured in large quantities by modern methods of peptide synthesis and then self-assembled into membranes using the layer-by-layer technique [24, 25]. The electrostatic layer-by-layer (LBL) assembly is driven primarily by coulombic interactions, but hydrophobic interactions and hydrogen bonds also contribute to film formation and stability, the amount depending on polypeptide design. Most peptides suitable for LBL assembly form films with a large percentage of beta-sheet at neutral pH; it would appear that beta-sheet is favored over alpha-helix in this context by the contribution to entropy of the number of ways of forming a beta-sheet from a single polypeptide chain. Film thickness and roughness depend rather substantially on amino acid composition [23].

A functional biomimetic membrane system will require the incorporation of membrane proteins to either have the role of an enzyme or to transport molecules or ions across the membrane. Hybrid systems that combine such biological proteins with synthetic membranes provide a practical solution for the creation of powerful diagnostic, preparative, and therapeutic devices. Indeed, artificial nanopores could also be used to replace biological proteins [26]. The use of artificial nanoprobes would allow completely synthetic systems to be constructed. The use of such biomimetic membrane systems, whether hybrid or synthetic, will require techniques to measure the transport of molecules across the biomimetic membrane. An elegant example of one such technique is optical single-transporter recording (OSTR), in which arrays of artificial micropores and nanopores are employed to analyze protein nanopores of cellular membranes. OSTR offers the possibility to be combined with electrical recording of ion transport or with 4Pi microscopy for single-molecule fluorescence spectroscopy and fluorescence correlation spectroscopy [27]. More recent research has developed this concept of such hybrid or synthetic membranes to have new properties (e.g., controlled permeability to ions/molecules), and functionality depending on the specificity of the inserted biomolecules [28]. The functional insertion of membrane proteins into synthetic membranes has provided appropriate properties to overcome the high hydrophobic mismatch between the size of the biomolecule and the membrane thickness. This approach has the promise of new solutions for the development of nano-reactors, artificial organelles or active surfaces/membranes that, by further gaining in complexity and functionality, will promote translational applications [28].

Nanoparticles with Biomimetic Membranes and Barrier Functions of Cell Membranes

Earlier in this chapter, we discussed the advances in the understanding of the biochemical and biophysical processes that have provided knowledge for the self-assembly of biomimetic membrane systems but have not provided knowledge of the underlying causal cellular component that imbues the essence of "life" for biomimetic membrane systems. We saw that essence would be of great value in overcoming the inability to reconstruct natural photosynthesis in an artificial biomimetic membrane system. Another example where that essence would be of value is for the engineering of biomimetic membrane systems that can mimic the uptake and diffusion of natural cells. If such a biomimetic system were to be successful, there would be an enormous value for such a system to be applied to medical diagnostic and therapeutic systems for nanoparticles. However, it seems the essence of "life" is a major missing link for that type of diagnostic biomimetic membrane system. This is seen in the elegant work of Banerji and Hayes (2007), who created phospholipid membrane systems to study the non-endocytic bulk transport of nanoparticles [29]. Those authors captured gold nanoparticle dispersions (7, 10, and 15 nm diameter particles) in giant unilamellar vesicles composed of phosphatidylcholine, phosphatidic acid, and cholesterol. Non-endocytotic transmembrane transport was modeled as the time derivative of the appearance of nanoparticles in the phosphate buffer outside the vesicles at 37 °C. The results show the transport rate to be zero, and, hence, a simple diffusive process of transmembrane transport is not supported.

A study published by He et al. (2016) on living MDCK epithelial cells demonstrated that dynamic processes are necessary for cellular uptake and transport of polymer nanoparticles [30]. First, the adhesion of polymer nanoparticles (PNs) on cell membranes was found to be time-dependent with a shift of both location and dispersion pattern, from the lateral adhesion of mainly mono-dispersed PNs initially to a later apical coverage of the PN aggregate. Then, it was interesting to observe in this study that the dynamic bio-adhesion of PNs only affected their endocytosis but not their transcytosis. It was important to find that the endocytosis of PNs was not a constant process. A GM1-dependent CDE (caveolae dependent endocytosis) pathway was dominant in the preliminary stage, followed by the co-existence of a CME (clathrin-mediated endocytosis) pathway for the PN aggregate at a later stage, in accordance with the adhesion features of PNs, suggesting the modification of PN adhesion patterns on the endocytosis pathways. Next, the PN adhesion was noticed to affect the structure of cell junctions, via altering the extra- and intra-cellular calcium levels, leading to the enhanced paracellular transport of small molecules, but not favorably enough for the obviously increased passing of PNs themselves. Finally, FRAP and other techniques all demonstrated the obvious impact of PN adhesion on the membrane confirmation, independent of the adhesion location and time, which might lower the threshold for the internalization of PNs, even their aggregates. Generally, these findings confirm that the transport pathway mechanism of PNs through epithelial cells is rather dynamic and is remarkably affected by the adhesion patterns of PNs on the cell membrane [30].

Contrasting those studies of Banerji and Hayes (2007) with He et al. (2016) allows us to draw three interesting insights for the medical diagnostic and therapeutic applications of biomimetic membrane systems. The first insight is that a complete biomimetic membrane system would require the elusive "essence of life" for that system to be truly a perfect medical diagnostic and therapeutic system that mimicked the precise functions of living cells and membrane systems. However, having said that, one can also draw the second insight that the current state-of-the-art biomimetic membrane system provides the ability to isolate the purely biophysical and biochemical bases for the function of biomimetic membranes. As with all areas of scientific research, the conclusions drawn from results must be analyzed with care and prudence.

The third insight is that a biomimetic membrane system can be usefully used as a medical diagnostic and therapeutic system by designing the system to place the biomimetic membrane in the correct position. For example, since it is clear that uptake and transport of nanoparticles is a biodynamic process that requires the "essence of life" in a cell, then the appropriate use of a biomimetic membrane system could be to coat the nanoparticle with a biomembrane system that could be designed with specific purposes such as enhanced biocompatibility, non-uptake by cells, or specific uptake by targeted cells.

We will elaborate on these insights in later chapters, but this third insight is exemplified by some recent publications concerning exosomes that put this part of the discussion into context, including, for example, the work of Plebanek et al [31]. Exosomes are nanoscale vesicles that mediate intercellular communication. Cellular exosome uptake mechanisms are not well defined partly due to the lack of specific inhibitors of this complex cellular process. Exosome uptake depends on cholesterol-rich membrane micro-domains called lipid rafts, and can be blocked by non-specific depletion of plasma membrane cholesterol. Scavenger receptor type B-1 (SR-B1), found in lipid rafts, is a receptor for cholesterol-rich high-density lipoproteins (HDL). In cell models, it was shown that HDL NPs bind SR-B1, activate cholesterol efflux, and attenuate the influx of esterified cholesterol. As a result, HDL NP treatment results in decreased dynamics and clustering of SR-B1 contained in lipid rafts and potently inhibits cellular exosome uptake. Thus, SR-B1 and targeted HDL NPs provide a fundamental advance in studying cholesterol-dependent cellular uptake mechanisms [31].

Those advances in understanding the cellular uptake mechanisms for exosomes, which are known to carry proteins and nucleic acids, make exosomes especially suitable for delivering biological therapeutics. However, the considerable complexity and the related high chance of off-target effects of these carriers are major barriers to translation to the clinic. Given the possibility that not all components of exosomes are required for their proper functioning, an alternative strategy would be to mimic these vesicles synthetically. Functional exosome mimetics may be created by the assembly of liposomes harboring only crucial components of natural exosomes. The low complexity and use of well-characterized components strongly increase the pharmaceutical acceptability of such systems. However, exosomal components that would be required for the assembly of functional exosome mimetics remain to be identified [32].

Those examples of preparing biomimetic exosomes that have similar targeting capabilities as natural exosomes also lead to a more synthetic example for biomimetic membrane technology. Several types of nanocarriers have been proposed for biomedical purposes, including inorganic nanoparticles, lipid aggregates, including liposomes, and synthetic polymeric systems, such as vesicles, micelles, or nanotubes. Polymeric vesicles, which are structures similar to lipid vesicles but created using synthetic block copolymers, are excellent candidates for new nanocarriers for medical applications. These structures are more stable than liposomes but retain their low immunogenicity. Significant efforts have been made to improve the size, membrane flexibility, and permeability of polymeric vesicles and to enhance their target specificity. Polymeric vesicles generated by the self-assembly of amphiphilic copolymers (polymersomes) offer the advantage of simultaneous encapsulation of hydrophilic compounds in their aqueous cavities and the insertion of fragile, hydrophobic compounds in their membranes. The insertion of membrane proteins or biopores into the polymer membrane can support the activity of co-encapsulated enzymes that act in tandem inside the cavity or of combinations of drugs and imaging agents. Also, surface functionalization of these nanocarriers permits specific targeting of the desired biological compartments. The combination of polymer vesicles and biological entities could provide tools to improve medical diagnosis and therapy [33].

Polymersomes are able to overcome the body's natural defences, remain stable for extended time in circulation, have tuneable membrane properties that allow the control of membrane permeability and therefore of drug release, and have the potential to be functionalized for active targeting of specific tissues to reduce undesirable side effects. Extensive work has been carried out in order to attain multifunctional polymeric vesicles that respond to precise triggers (e.g., temperature, pH, redox, and magnetic field) with spatial and temporal monitoring that may enable unprecedented control of drug release in the body. These versatile structures can be loaded with different types of (bio)molecules and nanoparticles, from drugs to contrast agents for medical imaging, and are able to accommodate them in different subcompartments of the vesicle (i.e., hydrophobic membrane and hydrophilic core). Multimodal targeted delivery systems could be obtained from this unique platform, with abilities in both drug delivery and medical imaging contrast enhancement, widening the perspectives toward theranostics. Polymersomes offer a promising route toward more effective treatments with fewer side effects and superior outcomes [34].

Overview of Advances in Biosensors and Nanobiotechnology

The concept of the lipid bilayer as providing a barrier function for the organization of cellular systems also provides a conceptual framework for bioinspired nanobiotechnology. That bioinspiration has two aspects. The first is that the lipid bilayer provides an electrochemically well-sealed barrier membrane to separate charged bioactive molecules into cellular compartments. The second is that proteins incorporated into the lipid bilayer provide an ordered flow of charged ions or an ordered sensing capability for external molecules that impinge onto the lipid bilayer. Both of those aspects are analogous to commonly used principles of electronics that are used to create sensor systems. However, nature has provided proteins for incorporation into the lipid bilayer that function as extremely sensitive nanostructured biosensors. In most cases, the natural protein biosensor has greater sensitivity and dynamic range with a much smaller size, as compared to a synthetic/artificial equivalent electronic component. Moreover, the intracellular second messenger signaling pathways provide an exquisitely controllable system for amplification of the binding of external molecules to membrane-bound protein receptors. These concepts will be discussed in greater detail in later chapters, particularly for the inclusion of a biomimetic membrane in a biosensor. In the remainder of this section, we summarize where some of the state of the art has reached in bioinspired biosensing systems.

For several years, there has been idea that the field of medically relevant technologies developed at the interface of life science and engineering will provide platform modalities that can be employed toward medical diagnostics and therapeutics [35]. Nonetheless, the devil is always in the detail to be able to develop those desired outcomes into practical technologies. An elegant approach to seeking the detail to be able to construct a bio-inspired biosensor is seen in the design of an electronic nose [36]. That electronic nose was constructed from polymers and multi-walled carbon nanotubes (MWCNT), but with an architecture that was designed to mimic the structure of the natural olfactory system in the nose. The mammalian olfactory receptor cells distributed in the olfactory epithelium are covered by a mucous layer, which functions to filter dust and impurities prior to a dissolved odorant being sensed by olfactory receptor cells. The targeted odors for that electronic nose were the volatiles from complex alcohols (e.g., sake, sorghum liquor, medical liquor, and whisky), and so the system was not a fully functional biomimetic membrane nose but rather a sensitive polymer-based gas sensor. Nonetheless, the design for that system was bioinspired by combining a thin polymer-capture layer (for the odorants) to coat a layer of MWCNT to provide the electrical conducting pathway. In effect, the polymer layer mimicked the mucous layer. For the "electronic nose" designed by Wang et al. (2011) there was a dramatic increase in sensitivity compared with traditional conducting polymer sensors that usually mix conducting materials (such as carbon black, CNTs, SWNTs, and MWNTs) and polymers into one sensing film layer. The other bioinspired aspect for the engineering of that "electronic nose" was to mimic the rapid response kinetics of typical biological action potentials carried by the olfactory nerve. Wang et al. (2011) devised a "bio-inspired fastadaptive readout circuit" to respond rapidly to the sensor response signal while tuning out long-term, constant signals from background humidity, temperature, and odors [36].

Although that type of bioinspired engineering approach can improve the sensitivity for more traditionally constructed biosensors, it does seem that the future directions for biosensor development will be better placed in the realm of biomimetic membrane technology. The basis for this observation is supported by the work in roadmapping the directions for biosensor research [37]. In that analysis, Siontorou and Batzias (2014) concluded that future trajectories for biosensor research and development are aligned with nanotechnology, and especially nanofabrication and nano-bioinformatics. Furthermore, the focus for such research and development would be aimed at designing biosensors that control the natural process of self-assembly and the thermodynamics of bioelement-lipid interaction. By that analysis, the future paradigm shifts in biosensors would be firmly in realm of biomimetic membrane technology and more generally in nanobiotechnology.

Biological Engineering and Nanobiotechnology

We can discuss such future biosensor paradigm shifts in terms of biological engineering. This is a concept we can introduce to describe the use of biological elements as components in the engineering of systems for nanobiotechnology applications. The term biological engineering thus goes beyond the idea of engineering systems by utilizing bioinspiration from nature. In Chap. 2, we will examine biological engineering from the perspective of engineering nanostructured systems with inspiration from biology.

The elegant feature of biology that encapsulates the concepts of biomimetics and bioinspiration is the self-assembly of molecules into functional systems at the dimensions of nanometers. Nonetheless, bioinspiration is usually interpreted as providing some biological-like characteristics that can be used to guide the synthesis of materials into functional devices and processes. Then, the result is the development of biotechnology systems that are based increasingly on the application of nanotechnology processes to create smaller objects. However, that interpretation overlooks the important concept that biology utilizes self-assembly of molecules to create elegant nanostructured systems. A central feature of such biological nanostructured systems is the assembly of phospholipid bilayer membranes that both provide compartments (i.e., biological cells) to rationalize the overall function of complex organisms (e.g., plants, animals) and also provide an environment in which to stabilize proteins to assist in the sensing and actuating functions of biological cells.

Thus, an alternative approach to developing biotechnology systems is through harnessing the bioinspired self-assembly of biological molecules to develop nanostructured biomimetic systems. Clearly, to make practical biotechnology systems does continue to require some aspects of materials synthesis, including polymers, metals, and ceramics. It follows that this approach to bioinspiration and biomimetic biotechnology will result in hybrid systems comprising synthetic materials that support and integrate self-assembled biological components.

In this chapter, we focus attention on discussing nanomedicine. This is to set the scene for the pragmatic approach of biological engineering that commences with the premise that the value of nanomedicine is to improve the health of individuals. This provides the "pull" from a clinical need to produce the solution that is to be addressed by multidisciplinary research and development from the biological engineering approach.

Nanomedicine and Nanobiotechnology

Advances in nanobiotechnology provide a rich resource for the development of nanomedicine. The European Science Foundation (ESF) defined the field of nanomedicine as the science and technology of diagnosing, treating, and preventing disease and traumatic injury, of relieving pain, and of preserving and improving human health by using molecular tools and molecular knowledge of the human body [38]. The ESF further described the field of nanomedicine as including the five main subdisciplines of *(i)* analytical tools, *(ii)* nanoimaging, *(iii)* nanomaterials and nanodevices, *(iv)* novel therapeutics and drug delivery systems, and *(v)* clinical, regulatory, and toxicological issues. In simpler terms, nanomedicine can be defined as the application of nanotechnology to medicine [39]. While various nanotechnologies continue to have an impact on nanomedicine, it is tempting to try to predict the emerging technologies that might have the most significant impact on nanomedicine in the coming years. Clearly, such predictions are fraught with difficulties. For example, in 2005 it was predicted that by 2015 to 2025 physicians would be using molecular machine systems and nanorobots, which would include an artificial mechanical red blood cell, an artificial mechanical white blood cell, or medical nanorobots capable of performing in vivo cytosurgery for chromosome replacement therapy [39]. Duncan (2005) took a more measured approach to the notion of nanorobots roaming the blood circulation to detect and treat disease [40]. However, such nanorobots have not yet entered routine medical practice. In 2015, a perspective on implications for health sciences suggested that, over the previous 10 years from 2005 to 2015, nanotechnologies have had a major role in the emergence of omics tools and in improving the modes of drug delivery [41]. Heath (2015) also pointed out that the fields of single-cell biology, as applied to assay biotechnologies, for example, and the harnessing of a patient's immune system for anti-cancer therapeutics were not anticipated in 2005. He also considered that the advances likely to occur in the decade 2015 to 2025 could include *(i)* nanotechnologies for health-related genomics and single-cell biology, *(ii)* inorganic and organic nanoparticles for biomedicine, and *(iii)* wearable nanotechnologies for wellness monitoring [41].

It is not really possible to arbitrarily provide pointers to the specific emerging technologies that may be applied to future developments in nanomedicine.

Nonetheless, here we can seek to provide a rational basis for reflecting on the possible ways in which emerging technologies could impact nanomedicine. To establish such a rational basis, we summarize the present state of the art in nanomedicine based on a bibliometric analysis of the existing published literature that is available in the Clarivate® Web of Science. We then build upon that state of the art to highlight the areas of nanomedicine where we believe that emerging technologies will be influential.

One of the key drivers for the emergence of novel technologies is to provide early detection of the molecular etiology of diseases in order to provide better means for timely interventions and to improve clinical outcomes. Examples of this include genome-based technologies and systems biology approaches that utilize information from the Human Genome Project (HGP). Next Generation Sequence (NGS) technology that developed as an outcome of the HGP has also become an important tool for generating information about chromatin structure and genome variations in terms of genetics and epigenetics. Molecular diagnosis can be made using NGS, and such genetic information can be used as a guide to potential therapy. With the recent development of CRISPR-Cas9 genome editing technology, gene therapies are becoming progressively more accessible. We can consider that there will be positions for biomimetic membrane nanobiotechnology, since it does seem that emerging technologies based on bioinspired and biomimetic approaches will provide important future directions for applications in nanomedicine.

State of the Art in Nanomedicine

The state of the art of nanobiotechnologies applied to the field of nanomedicine was surveyed within the time-period window of 1900 to 2021 by searching the published literature that is available in the Clarivate® Web of Science. To focus on specific aspects of the various nanobiotechnologies, the published literature was searched using various combinations of the keywords *"biomedicine," "nanomedicine," "biomimetic," "bioinspiration," "disease," "therapy," "human," "microbiota," "biotechnology," "nanobiotechnology," "nanosensor,"* and *"membrane."* The details of the searches are presented in Table 1.1. All of the BIOINSP, THERAP, MICROB, MEMBR, BILAYER, NANSEN, and NANBIO literature searches were analyzed using the "Bibliometrix" library of RStudio v1.2.5033 [42]. We consider the organization of these published documents within the broad categories of clinical and technological. Within the clinical category, we will group the searches BIONSP, THERAP, and MICROB. Within the technological category, we will group the searches MEMBR, BILAYER, NANSEN, and NANBIO.

The main descriptive data from the searches are summarized in Table 1.2. The field of nanomedicine is large, particularly when trying to identify nanobiotechnologies that are pertinent to both the state-of-the-art and to future directions in the field of nanomedicine. We have used the bibliometrix analyses to investigate the conceptual structure of the published state of the art to determine the main themes

Table 1.1 Terms used for the searches of the published literature available in the Clarivate® Web of Science Core Collection. In the descriptions of the searches, the field tags are either ALL = All fields or TS = Topic. The searches use the Boolean operators AND and OR. The asterisk (*) is used as the wildcard character. The total number of published documents resulting from each search is given in the column headed N

Search	Terms	N
BIONSP	((TS = (biomedicine)) OR TS = (nanomedicine)) AND ((TS = (biomimet*) OR (TS = (bioinspir*))))	616
THERAP	(ALL = (((biomedicine OR nanomedicine) AND (biomimet* OR bioinspir*)) AND (disease OR therap*)))	568
MICROB	((ALL = ((medicine AND (disease OR therap*)) AND human AND (microbiom* OR microbiot*))) AND (ALL = (biotechnolog*)))	532
MEMBR	((ALL = ((nanomedicine OR implant* OR biocompat*) AND (disease OR therap*) AND human AND biotechnolog*)) AND (ALL = (membrane)))	119
NANBIO	((ALL = (nanobiotechnolog* OR bionanotechnolog*)) AND (ALL = (biomimet* OR bioinspir*)))	345
BILAYER	(((ALL = (lipid bilayer)) AND (ALL = (biomimet* OR bioinspir*))) AND (ALL = (nanotechnolog*)))	50
NANSEN	((ALL = (biomimet* OR bioinspir*)) AND (ALL = (nanosens*)))	118

and trends. For that target of the Bibliometrix analyses, it is less important to investigate the intellectual structure (i.e., the influence of particular authors) or the social structure (i.e., the interaction of authors, institutions, or countries) of the published state of the art.

The next sections will describe the details of each of these bibliometrix analyses and then summarize the common aspects of the published state of the art in the field of nanomedicine. Thus, to investigate the conceptual structure, we focus on keywords rather than authors, institutions, countries, or co-citations. The focus on keywords is chosen to consider the trends of topics and themes of published state-of-the-art research, and not on the personalities nor countries from where the published research originated. Then at the end of this chapter, based on those bibliometrix analyses and summaries, we will seek to foresee future directions in the field of nanomedicine. Such foresight into future directions will seek not to reproduce the existing state of the art in the field of nanomedicine, but rather to examine the directions in which the potential for novelty and originality appears to be heading.

Bibliometrix Analysis of Nanomedicine Literature: Clinical

This aspect of the analysis utilizes the literature found from the searches BIONSP, THERAP, and MICROB. We commence by examining the most frequent word terminology used by authors within each area of the literature (see Table 1.1). These most frequently used words are represented in the wordclouds presented in Fig. 1.1. We can use these wordclouds to provide qualitative overview of the popular

Table 1.2 Main descriptive results from the Bibliometrix analyses for each of the searches described in Table 1.1. The types of documents are listed separately. The results for the number of authors per document and the average citations per document are referred to as the total number of documents

Description	BIONSP	THERAP	MICROB	MEMBR	NANBIO	BILAYER	NANSEN
Total number of documents	616	568	532	119	345	50	118
Article	371	365	376	92	212	35	62
Article, book chapter	13	9	1	1	9	3	7
Article, proceedings paper	6	7	3	1	–	–	7
Article, data paper	–	–	1	–	–	–	–
Correction	1	–	–	–	–	–	–
Editorial material	2	3	6	–	3	–	–
Letter	–	–	2	–	–	–	–
News item	–	–	–	–	1	–	1
Meeting abstract	3	–	1	–	–	–	2
Proceedings paper	13	9	3	–	13	2	29
Review	204	171	138	25	91	10	9
Review, book chapter	3	4	1	–	–	–	1
Total number of authors	2649	2565	6351	799	1404	219	540
Number of authors per document	4.3	4.5	11.9	6.7	4.1	4.4	4.6
Average citations per document	31.6	30.4	71.7	28.0	38.9	27.8	19.9
Keywords Plus	2383	2157	2160	724	1557	320	511

concepts in these areas of the literature. In those wordclouds, the size of the letters in the words represents the frequency of occurrence of the words. Thus, the most commonly used words are written in letters with the largest sizes.

It is not surprising that the wordclouds are similar for the BIONSP and THERAP literature. Those two areas of literature were searched using similar keywords but formulated in different ways for the search (Table 1.1). Although one could consider that microbiota is an essential part of medicine, authors publishing the documents in the field identified using the MICROB search have not included nanomedicine as a common word in their publications (Fig. 1.1c). That probably reflects that this literature has a stronger basis in microbiology and bacteria. Across those three

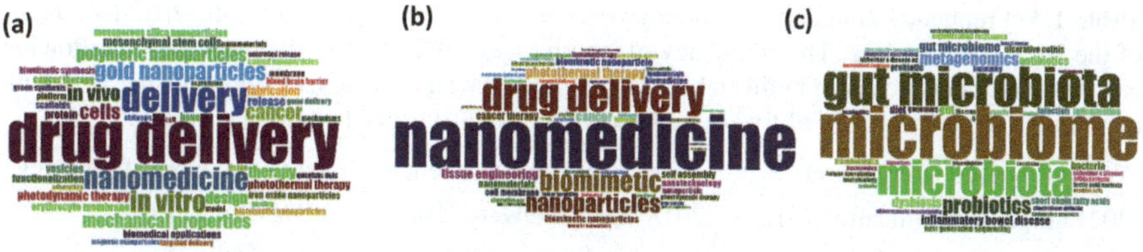

Fig. 1.1 Wordclouds of the most frequent words returned from the clinically oriented searches. (**a**) Wordcloud returned by the Bibliometrix analysis of search BIONSP. (**b**) Wordcloud returned by the Bibliometrix analysis of search THERAP. (**c**) Wordcloud returned by the Bibliometrix analysis of search MICROB

wordclouds of Fig. 1.1, it is not surprising that the most commonly used words in the clinically oriented literature are the words *drug delivery*, *nanomedicine,* and *microbiome*. We can seek more detailed and quantitative insights by applying further bibliometrix analyses of the conceptual structure to gain insights into the potential directions for the future of nanobiotechnology and nanomedicine.

Conceptual Structure of BIONSP Literature

The BIONSP search was conducted to focus on research concerning the diseases and therapeutic aspects of nanomedicine related to biomimetics or bioinspiration using the terms *((TS = (biomedicine)) OR TS = (nanomedicine)) AND ((TS = (biomimet*) OR (TS = (bioinspir*))))*. These terms retrieved 616 documents (Table 1.2). We first quantify the number of times a keyword was used in each year, with the threshold for the minimum frequency set at five times in any year (Table 1.3). In the period from 2010 to 2021, there are 36 keywords that satisfy that usage restriction. Of those keywords, there are 12 identifiable as technically oriented and 21 identifiable as biologically oriented, with 3 keywords that could be in either orientation (*mechanisms*, *adsorption*, and *adhesion*). The most frequently used keywords were *nanoparticles* (118 times in 2019) and *drug delivery* (106 times in 2018).

The most frequently used keywords for each year are shown in Fig. 1.2. During the period 2013 to 2018, the most frequently used keyword was biologically oriented, whereas since 2019 this has been technically oriented.

Then we quantify the co-occurrence of keywords identified in the searches shown in Table 1.1. This co-occurrence analysis helps to identify the commonality of the threads of research that have been published and is shown in Fig. 1.3 as a co-occurrence network for the literature identified in BIONSP. The co-occurrence network is shown as a Kamada & Kawai network layout [43]. In this network, each keyword is represented as a node that has a size proportional to the frequency of the keyword's occurrence. The lines represent the connections between nodes, and the line thickness reflects the extent of co-citation between nodes. The weight of the link connecting the pair of keywords quantifies the number of times that a pair of

Table 1.3 Frequency of occurrence of keywords for each year of publication for 616 documents of the BIONSP literature. The frequency of occurrence is indicated in the parentheses following each keyword. For inclusion in this table, we restricted the analysis to five keywords per year with a minimum word frequency of five (e.g., no keywords in 2011 matched these criteria)

Year	Keyword (frequency)	Year	Keyword (frequency)	Year	Keyword (frequency)
2021	Signal amplification (5)	2018	Drug delivery (106)	2016	Bone (17)
2020	Mesoporous silica nanoparticles (17)		Delivery (65)		Cell (12)
	Mechanisms (13)		Nanomedicine (48)		Adsorption (12)
	Membrane-coated nanoparticles (10)		In vitro (46)		Particles (11)
	Dendritic cells (10)		Gold nanoparticles (38)		Composite scaffolds (10)
	Differentiation (9)	2017	Vesicles (19)		Carbon nanotubes (10)
2019	Nanoparticles (118)		Biomedical applications (17)		Adhesion (10)
	Cells (35)		Biomimetic synthesis (15)	2015	Cytoxicity (5)
	Cancer (34)		Gene delivery (14)	2014	Endothelial cells (9)
	Polymeric nanoparticles (29)		Tissue (13)		Physical properties (5)
	Therapy (26)		Quantum dots (13)	2013	Recognition (5)
			Chitosan (13)	2012	Surfaces (6)
				2010	Biocompatibility (5)

keywords co-occurs in multiple documents. These co-occurring keywords cluster into four communities, which are identified by the colors purple (cluster 1), green (cluster 2), blue (cluster 3), and red (cluster 4). These communities of clustered co-occurring keywords are interpreted as representing three main types of research, which include *fundamental biology* (purple cluster 1), *clinical biology* (green cluster 2), and *therapeutic applications* (blue cluster 3, and red cluster 4). The keywords at the center of the purple *(drug delivery)* and green *(nanoparticles)* clusters represent the strongest co-occurring keywords in the BIONSP documents.

Table 1.4 summarizes the keyword-nodes that are clustered closest to the central themes of the communities of the documents identified in the BIOINSP search. There are four central themes identified as the node with the largest index of centrality, W, including *drug delivery* (fundamental biology), *nanoparticles* (clinical biology), in vivo (therapeutic applications), and *polymeric nanoparticles* (therapeutic applications).

We apply similar analyses to the literature identified by the THERAP search. The THERAP search included additional and specific terms that were aimed to identify documents that were specifically related to either therapy or diagnosis.

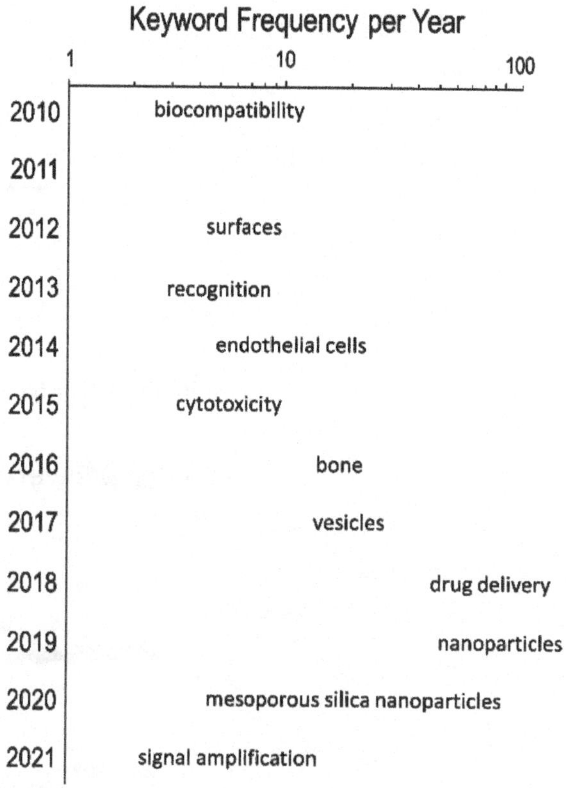

Fig. 1.2 Most frequently used keywords in each year of publication for 616 documents of the BIONSP literature. This provides some insight into the trending topics in the BIONSP literature. The keywords in this figure are the most frequently used keywords in each year from the list shown in Table 1.3

Conceptual Structure of THERAP Literature

The THERAP search was conducted to focus on research concerning the diseases and therapeutic aspects of nanomedicine using the terms *(ALL = (((biomedicine OR nanomedicine) AND (biomimet* OR bioinspir*)) AND (disease OR therap*)))*. These terms retrieved 568 documents (Table 1.2). We first quantify the number of times a keyword was used in each year, with the threshold for the minimum frequency set at five times in any year (Table 1.5). In the period from 2010 to 2021 there are 37 keywords that satisfy that usage restriction. Of those keywords, there are 10 identifiable as technically oriented and 26 identifiable as biologically oriented, with 1 keyword that could be in either orientation (*adsorption*). The most frequently used keywords were *nanoparticles* (125 times in 2019) and *drug delivery* (122 times in 2018). It is not surprising that the keyword frequency identified in the THERAP literature is similar to that in the BIONSP literature.

The most frequently used keywords for each year are shown in Fig. 1.4. Since 2012, there has been a reasonably equal mix of technically oriented (60%) and biologically oriented keywords (40%).

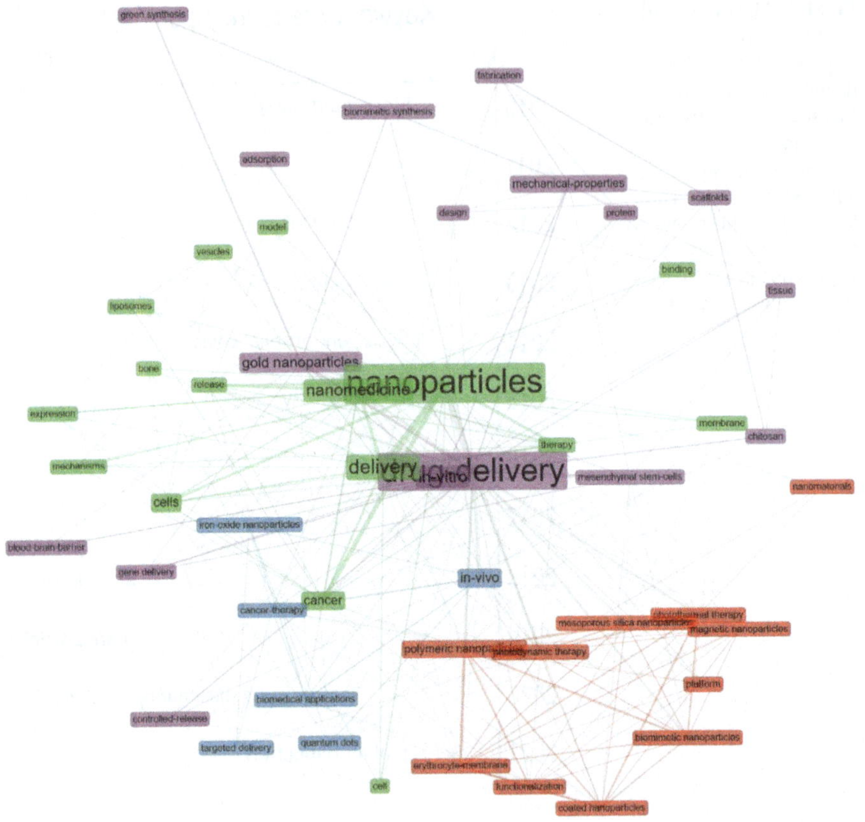

Fig. 1.3 Co-occurrence network of keywords from the Bibliometrix analysis of search BIONSP. The co-occurrence associations for those keywords are shown as a Kamada & Kawai network layout with normalization by association and using the Louvain clustering algorithm. The layout is shown for 50 nodes

Then we quantify the co-occurrence of keywords identified in the searches shown in Table 1.1. This co-occurrence analysis helps to identify the commonality of the threads of research that have been published and is shown in Fig. 1.5 as a co-occurrence network for the literature identified in THERAP. The co-occurrence network is shown as a Kamada & Kawai network layout [43]. In this network, each keyword is represented as a node that has a size proportional to the frequency of the keyword's occurrence. The lines represent the connections between nodes, and the line thickness reflects the extent of co-citation between nodes. The weight of the link connecting the pair of keywords quantifies the number of times that a pair of keywords co-occurs in multiple documents. These co-occurring keywords cluster into three communities, which are identified by the colors red (cluster 1), blue (cluster 2), and green (cluster 3). These communities of clustered co-occurring keywords are interpreted as representing three main types of research, which include *fundamental biology* (red cluster 1), *clinical biology* (blue cluster 2), and *therapeutic*

Table 1.4 Co-occurrence of keywords from BIOINSP listed in clusters according to the colors shown in Fig. 1.3. A keyword with a larger index of centrality (W) is positioned more centrally to the cluster. A threshold for co-occurrence clustering was set for W > 5 (keywords shaded in grey)

Cluster 1 (purple) Fundamental biology (W)	Cluster 2 (green) Clinical biology (W)	Cluster 3 (blue) Therapeutic applications (W)	Cluster 4 (red) Therapeutic applications (W)
drug delivery (426.5)	nanoparticles (201.7)	in vivo (13.8)	polymeric nanoparticles (24.4)
gold nanoparticles (55.4)	delivery (55.8)	iron-oxide nanoparticles (4.9)	photodynamic therapy (9.4)
in vitro (29.3)	nanomedicine (32.9)	biomedical applications (4.5)	erythrocyte membrane (6.1)
mechanical properties (9.5)	cancer (18.0)	cancer therapy (4.2)	mesoporous silica nanoparticles (5.3)
biomimetic synthesis (6.2)	therapy (7.4)	targeted delivery (1.9)	photothermal therapy (5.2)
mesenchymal stem cells (6.0)	cells (7.3)	quantum dots (1.8)	magnetic nanoparticles (3.0)
protein (2.6)	release (3.3)		biomimetic nanoparticles (2.7)
scaffolds (2.5)	liposomes (1.2)		functionalization (2.4)
design (2.3)	cell (1.1)		coated nanoparticles (1.5)
controlled release (0.8)	vesicles (1.0)		platform (0.2)
chitosan (0.8)	bone (0.6)		
tissue (0.8)	binding (0.3)		
fabrication (0.6)	mechanisms (0.2)		
gene delivery (0.5)	model (0.2)		
blood brain barrier (0.4)	membrane (0.2)		
	expression (0.2)		

applications (green cluster 3). The keywords at the center of the red *(drug delivery)* and blue *(nanoparticles)* clusters represent the strongest co-occurring keywords in the THERAP documents.

Table 1.6 summarizes the keyword nodes that are clustered closest to the central themes of the communities of the documents identified in the THERAP search. There are three central themes identified as the node with the largest index of centrality, W, including *drug delivery* (fundamental biology), *nanoparticles* (clinical biology), and *polymeric nanoparticles* (therapeutic applications).

Conceptual Structure of MICROB Literature

The MICROB search was conducted to focus on research concerning the microbiota using the terms *((ALL = ((medicine AND (disease OR therap*)) AND human AND (microbiom* OR microbiot*))) AND (ALL = (biotechnolog*)))*. These terms retrieved 532 documents (Table 1.2). We first quantify the number of times a keyword was used in each year, with the threshold for the minimum frequency set at five times in any year (Table 1.7). In the period from 2012 to 2020, there are 31 keywords that satisfy that usage restriction. All of those keywords there are identifiable as biologically oriented. The most frequently used keywords were *microbiota* (46 times in 2019) and *gut microbiota* (89 times in 2018).

The most frequently used keyword for each year are shown in Fig. 1.6.

Then we quantify the co-occurrence of keywords identified in the searches shown in Table 1.1. This co-occurrence analysis helps to identify the commonality of the threads of research that have been published and is shown in Fig. 1.7 and as a co-occurrence network for the literature identified in MICROB. The co-occurrence network is shown as a Kamada & Kawai network layout [43]. In this network, each

Table 1.5 Frequency of occurrence of keywords for each year of publication for 568 documents of the THERAP literature. The frequency of occurrence is indicated in the parentheses following each keyword. For inclusion in this table, we restricted the analysis to five keywords per year with a minimum word frequency of five

Year	Keyword (frequency)	Year	Keyword (frequency)	Year	Keyword (frequency)
2021	Hypoxia (5)	2018	Drug delivery (122)	2015	Transport (9)
2020	Mesoporous silica nanoparticles (17)		In vitro (55)		Adsorption (9)
	Liposomes (17)		Therapy (48)		Phase I (7)
	Tumor (14)		In vivo (48)		Cancer cells (6)
	Resistance (14)		Nanomedicine (48)	2014	Nanocrystals (8)
	Dendritic cells (13)	2017	Nanocarriers (20)	2013	Fabrication (9)
2019	Nanoparticles (125)		Iron oxide nanoparticles (17)		Particles (8)
	Delivery (73)		Peptide (14)		Bone (6)
	Cancer (40)		Vesicles (13)		Macromolecular therapeutics (5)
	Polymeric nanoparticles (35)		Nanomaterials (13)	2012	Mechanical properties (12)
	Mesenchymal stem cells (28)	2016	Gene delivery (17)		Biocompatibility (5)
			Tissue (16)		
			Scaffolds (15)		
			Extracellular matrix (13)		
			Biomedical applications (13)		

keyword is represented as a node that has a size proportional to the frequency of the keyword's occurrence. The lines represent the connections between nodes, and the line thickness reflects the extent of co-citation between nodes. The weight of the link connecting the pair of keywords quantifies the number of times that a pair of keywords co-occurs in multiple documents. These co-occurring keywords cluster into four communities, which are identified by the colors purple (cluster 1), red (cluster 2), blue (cluster 3), and green (cluster 4). These communities of clustered co-occurring keywords are all related closely to theme of microbiology and the gut microbiota. Nonetheless, we can interpret these as representing three axes of research within the theme, which include *microbiota* (purple cluster 1 and red cluster 2), *bacterial* (blue cluster 3), and *microbiology* (green cluster 4). The keywords at the center of the purple *(gut microbiota)* and blue *(diversity)* clusters represent the strongest co-occurring keywords in the MICROB documents.

Table 1.8 summarizes the keyword nodes that are clustered closest to the central themes of the communities of the documents identified in the MICROB search. There are four central themes identified as the node with the largest index of centrality, W, including *gut microbiota* (microbiota), *microbiota* (microbiota), *diversity* (bacterial), and *expression* (microbiology).

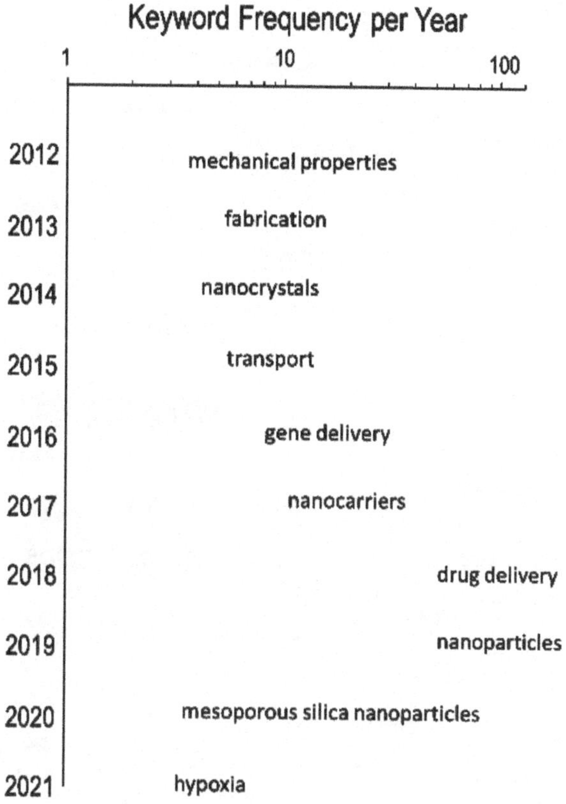

Fig. 1.4 Most frequently used keywords for each year of publication for 616 documents of the THERAP literature. These words are the most frequently used keywords from the list shown in Table 6, and provide some insight into the trending topics in the THERAP literature

Summary of the Conceptual Structure of the Nanomedicine Literature: Clinical

We have investigated the conceptual structure of the nanomedicine literature using three separate searches, followed by a bibliometrix analysis. The search terms in those literature searches were related but also included different terms to investigate the clinical nanomedicine literature over a wider context. We can consider that the conceptual structure of the clinical nanomedicine literature is clustered around three central themes of research, which are *fundamental biology*, *clinical biology*, and *therapeutic applications*. Research concerning the microbiota can be considered as sub-themes of fundamental and clinical biology. Within those clustering themes, the dominant keyword nodes were identified as those with indices of centrality greater than 200. These dominant keyword nodes are *drug delivery*, *nanoparticles*, and *gut microbiota*.

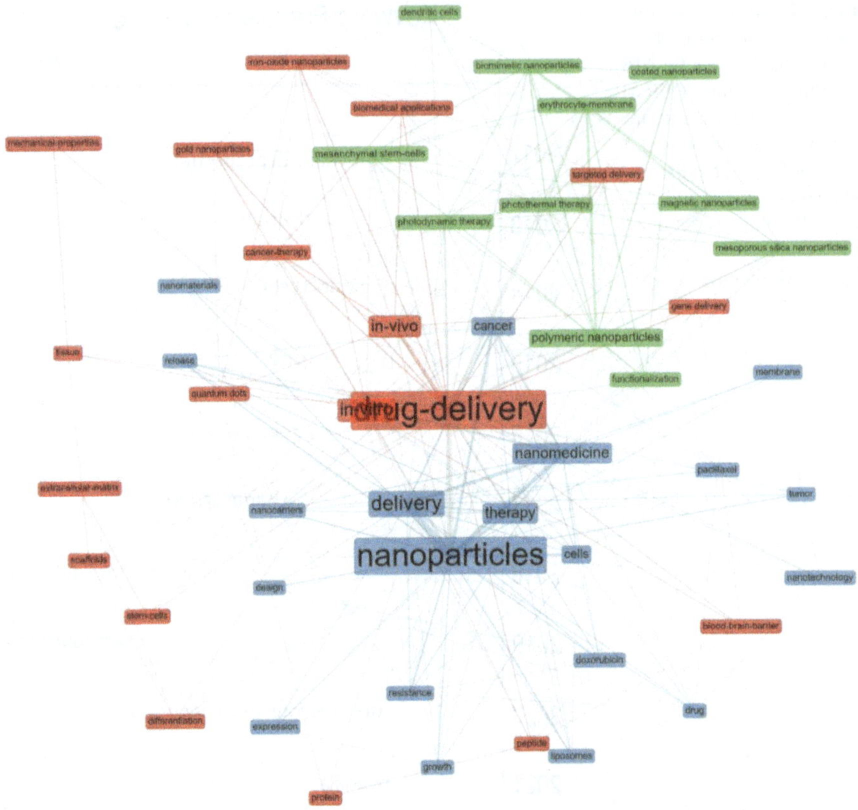

Fig. 1.5 Co-occurrence network of keywords from the Bibliometrix analysis of search THERAP. The co-occurrence associations for those keywords are shown as a Kamada & Kawai network layout with normalization by association and using the Louvain clustering algorithm. The layout is shown for 50 nodes

Bibliometrix Analysis of Nanomedicine Literature: Technological

This aspect of the analysis utilizes the literature found from the searches MEMBR, BILAYER, NANSEN, and NANBIO. We commence by examining the most frequent word terminology used by authors within each area of the literature (see Table 1.1). These most frequently used words are represented in the wordclouds presented in Fig. 1.8. We can use these wordclouds to provide qualitative overview of the popular concepts in these areas of the literature. In those wordclouds, the size of the letters in the words represents the frequency of occurrence of the words. Thus, the most commonly used words are written in letters with the largest sizes.

These searches applied terms that were aimed to restrict the retrieved documents to focus on the technological aspects of membranes, nanobiotechnology, lipid

Table 1.6 Co-occurrence of keywords from THERAP listed in clusters according to the colors shown in Fig. 1.5. A keyword with a larger index of centrality (W) is positioned more centrally to the cluster. A threshold for co-occurrence clustering was set for W > 5 (keywords shaded in grey)

Cluster 1 (red)	Cluster 2 (blue)	Cluster 3 (green)
Fundamental biology (W)	Clinical biology (W)	Therapeutic applications (W)
drug delivery (342.9)	nanoparticles (149.5)	polymeric nanoparticles (37.4)
in vitro (82.8)	delivery (69.8)	mesenchymal stem cells (18.9)
in vivo (24.0)	therapy (27.1)	photodynamic therapy (11.2)
tissue (3.3)	nanomedicine (24.2)	functionalization (8.1)
cancer therapy (2.7)	cancer (12.9)	photothermal therapy (8.0)
iron oxide nanoparticles (2.1)	cells (8.0)	erythrocyte membrane (6.8)
Peptide (1.6)	design (2.7)	mesoporous silica nanoparticles (2.8)
quantum dots (1.6)	doxorubicin (1.8)	biomimetic nanoparticles (2.6)
Biomedical applications (1.5)	drug (1.6)	coated nanoparticles (1.0)
gold nanoparticles (1.4)	release (1.7)	magnetic nanoparticles (0.7)
gene delivery (1.0)	nanocarriers (1.5)	dendritic cells (0.3)
targeted delivery (0.9)	paclitaxel (0.9)	
scaffolds (0.6)	liposomes (0.7)	
extracellular matrix (0.4)	membrane (0.5)	
blood brain barrier (0.4)	resistance (0.4)	
protein (0.3)	growth (0.2)	
differentiation (0.3)	nanotechnology (0.2)	
stem cells (0.2)	nanomaterials (0.1)	
	tumor (0.1)	
	expression (0.1)	

bilayers and nanosensors. The wordclouds suggest that this focus was achieved, except for the MEMBR search, where only *amniotic membrane* appeared in the wordcloud as one of the more frequently used words. This initial observation suggested that the technological usage of membranes, such as biomimetic membrane technology, has not been a major theme of the nanomedicine literature to date. We can seek more detailed and quantitative insights by applying further bibliometrix analyses of the conceptual structure to gain insights into the potential directions for the future of biomimetic membrane technology, nanobiotechnology, and other technological aspects applied in the field of nanomedicine.

Conceptual Structure of MEMBR Literature

The MEMBR search was conducted to focus on membrane-related aspects of nanomedicine using the terms *((ALL = ((nanomedicine OR implant* OR biocompat*) AND (disease OR therap*) AND human AND biotechnolog*)) AND (ALL = (membrane)))*. These terms retrieved 119 documents (Table 1.2). We first quantify the number of times a keyword was used in each year, with the threshold for the minimum frequency set at five times in any year (Table 1.9). In the period from 2014 to 2021, there are 18 keywords that satisfy that usage restriction. Of those keywords,

Table 1.7 Frequency of occurrence of keywords for each year of publication for 532 documents of the MICROB literature. The frequency of occurrence is indicated in the parentheses following each keyword. For inclusion in this table, we restricted the analysis to five keywords per year with a minimum word frequency of five (e.g., no keywords in 2011 matched these criteria)

Year	Keyword (frequency)	Year	Keyword (frequency)	Year	Keyword (frequency)
2020	Activation (16)	2018	Gut microbiota (89)	2016	Fecal microbiota (26)
	Insulin resistance (12)		Diversity (49)		Crohn's disease (20)
	Colitis (11)		Bacteria (48)		Sequences (11)
	Growth (10)		Expression (39)		DNA (8)
	Infections (10)		Intestinal microbiota (36)		In vivo (8)
	Cancer (10)	2017	Disease (39)	2015	Irritable bowel syndrome (8)
2019	Microbiota (46)		Inflammatory bowel disease (24)		Populations (8)
	Inflammation (31)		Double blind (16)	2012	Microflora (7)
	Cells (31)		Ulcerative colitis (14)		
	Infection (25)		Induction (10)		
	Escherichia coli (22)		Probiotics (10)		
			Acid (10)		

there is 1 identifiable as technically oriented and 16 identifiable as biologically oriented, with 1 keyword that could be in either orientation (*mechanisms*, *adsorption*, and *adhesion*). The most frequently used keywords were *delivery* (15 times in 2017) and *cells* (14 times in 2014). This dominance by biologically oriented keywords is surprising since the search terms were more restrictive toward biotechnology and membranes, as compared to the most comparable clinically oriented search, THERAP.

The most frequently used keywords for each year are shown in Fig. 1.9.

The frequency of keyword usage suggests that the conceptual structure for the MEMBR literature remains dominated by clinical themes, with the keyword *membrane* most likely used in reference to cells.

We can further investigate the conceptual structure by quantifying the co-occurrence of keywords identified in the MEMBR search. This co-occurrence analysis helps to identify any commonality to the threads of research that have been published and is shown in Fig. 1.10 as a Kamada & Kawai network layout [43]. In this network, each keyword is represented as a node that has a size proportional to the frequency of the keyword's occurrence. The lines represent the connections between nodes, and the line thickness reflects the extent of co-citation between nodes. The weight of the link connecting the pair of keywords quantifies the number of times that a pair of keywords co-occurs in multiple documents. These co-occurring keywords cluster into five communities, which are identified by the colors orange (cluster 1), blue (cluster 2), violet (cluster 3), green (cluster 4), and yellow (cluster 5).

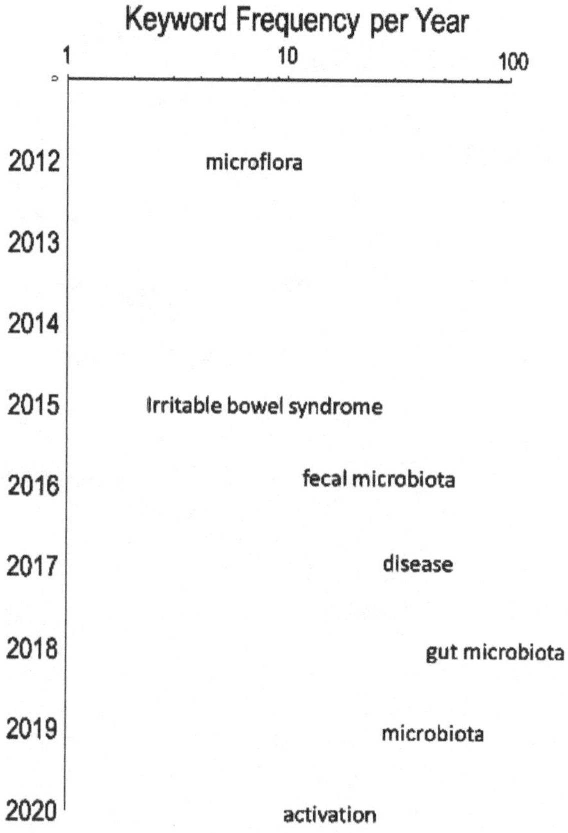

Fig. 1.6 Most frequently used keywords for each year of publication for 532 documents of the MICROB literature. These words are the most frequently used keywords from the list shown in Table 8, and provide some insight into the trending topics in the MICROB literature

Note that several keywords have zero indices of centrality and are displayed around the periphery of the network. Such a distribution suggests that those keywords would point to emerging themes of research that are within the conceptual framework defined by the terms used for the MEMBR search. Nonetheless, some of those non-central keywords have some association with the central clusters, indicated by a similar color for the node. Those non-central keywords are *resistance* (violet), *proliferation* (violet), *gold nanoparticles* (blue), *membrane* (green), *grafts* (orange), *apoptosis*, *performance*, and *clinical applications*. Even though the keyword *membrane* is clustered centrally (W = 125.3) within the clinical biology community (violet), it also appears as one of the non-central keywords in a different color for the node. This suggests that authors have utilized the keyword *membranes* in a context that does not co-occur within a clinical biology cluster. Such a context would most likely be the use of membranes in a technological context and would provide evidence for an emerging new research theme, such as biomimetic membrane biotechnology.

The communities of clustered co-occurring keywords are all related closely to the general theme of biology and can be interpreted as representing three axes of

Fig. 1.7 Co-occurrence network of keywords from the Bibliometrix analysis of search MICROB. The co-occurrence associations for those keywords are shown as a Kamada & Kawai network layout with normalisation by association and using the Louvain clustering algorithm. The layout is shown for 50 nodes

research within the theme, which include *fundamental biology* (orange cluster 1), *clinical biology* (blue cluster 2, violet cluster 3, and green cluster 4), and *ocular biology* (yellow cluster 5). The keywords at the center of the orange (in vitro), blue (*cells*), violet (*transplantation*), and green *(controlled release)* clusters represent the strongest co-occurring keywords in the MEMBR documents (W > 100, Table 1.10).

Table 1.10 summarizes the keyword nodes that are clustered closest to the central themes of the communities of the documents identified in the MEMBR search. There are five central themes identified as the node with the largest index of centrality, W, including in vitro (fundamental biology), *cells* (clinical biology), *transplantation* (clinical biology), *controlled release* (clinical biology), and *Bruchs membrane* (ocular biology).

Table 1.8 Co-occurrence of keywords from MICROB listed in clusters according to the colors shown in Fig. 1.7. A keyword with a larger index of centrality (W) is positioned more centrally to the cluster. A threshold for co-occurrence clustering was set for W > 5 (keywords shaded in grey)

Cluster 1 (purple)	Cluster 2 (red)	Cluster 3 (blue)	Cluster 4 (green)
Microbiota (W)	Microbiota (W)	Bacterial (W)	Microbiology (W)
gut microbiota (317.0)	microbiota (77.9)	diversity (141.0)	expression (56.0)
bacteria (65.9)	disease (43.5)	intestinal microbiota (84.5)	inflammation (29.3)
cells (26.8)	metabolism (25.4)	fecal microbiota (24.2)	chain fatty acids (14.8)
health (14.5)	gut (11.9)	diet (11.1)	mice (4.6)
inflammatory bowel disease (6.2)	risk (8.8)	identification (9.9)	activation (3.3)
Escherichia coli (5.4)	infection (8.3)	Crohns disease (6.8)	ulcerative colitis (1.6)
impact (1.3)	children (8.2)	obesity (3.3)	in vitro (0.6)
colitis (0.7)	mechanisms (7.3)	evolution (1.2)	
gene expression (0.3)	acid (4.0)	sequences (0.9)	
association (0.2)	double-blind (3.2)	colonization (0.8)	
	gene (3.0)	microbiome (0.5)	
	resistance (1.4	human gut microbiome (0.4)	
	prevalence (1.1)	reveals (0.3)	
	cardiovascular disease (0.7		

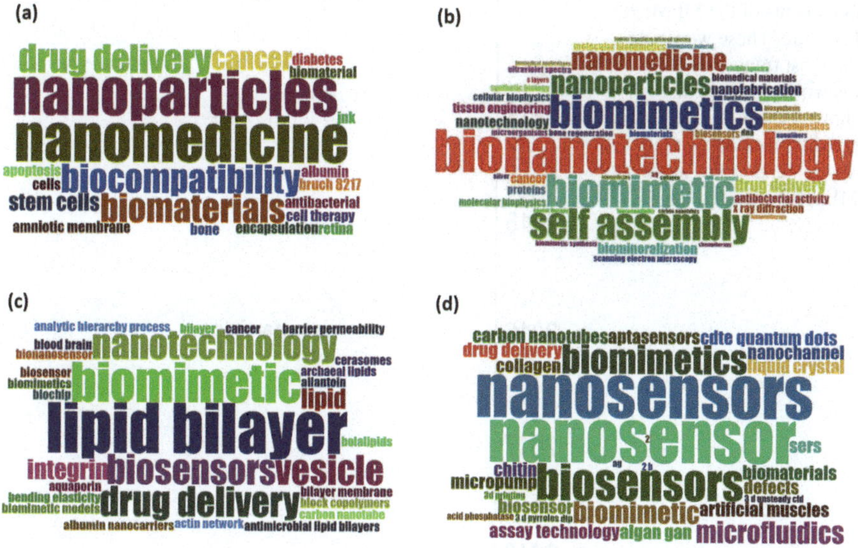

Fig. 1.8 Wordclouds of the most frequent words returned from the technically oriented searches. (**a**) Wordcloud returned by the Bibliometrix analysis of search MEMBR. (**b**) Wordcloud returned by the Bibliometrix analysis of search NANBIO. (**c**) Wordcloud returned by the Bibliometrix analysis of search BILAYER. (**d**) Wordcloud returned by the Bibliometrix analysis of search NANSEN

Conceptual Structure of NANBIO Literature

The NANBIO search was conducted to focus on nanobiotechnology using the terms *((ALL = (nanobiotechnolog* OR bionanotechnolog*)) AND (ALL = (biomimet* OR bioinspir*)))*. These terms retrieved 345 documents (Table 1.2). We first

Table 1.9 Frequency of occurrence of keywords for each year of publication for 119 documents of the MEMBR literature. The frequency of occurrence is indicated in the parentheses following each keyword. For inclusion in this table, we restricted the analysis to five keywords per year with a minimum word frequency of five (e.g., no keywords in 2011 matched these criteria)

Year	Keyword (frequency)	Year	Keyword (frequency)	Year	Keyword (frequency)
2020	Mesenchymal stem cells (9)	2017	Delivery (15)	2016	Nanoparticles (9)
2018	In vitro (13)		Expression (13)		Therapy (7)
	Drug delivery (10)		Transplantation (11)		Scaffolds (6)
	Stem cells (9)		Differentiation (11)		Extracellular matrix (6)
	Membrane (9)		Controlled release (6)	2015	In vivo (8)
	Angiogenesis (6)			2014	Cells (14)
					Biocompatibility (7)

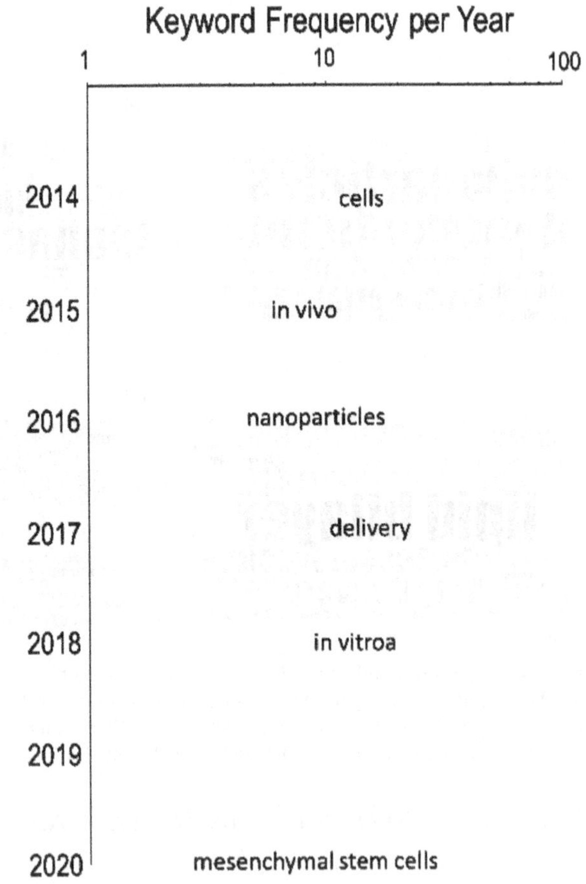

Fig. 1.9 Most frequently used keywords for each year of publication for 119 documents of the MEMBR literature. These words are the most frequently used keywords from the list shown in Table 1.9, and provide some insight into the trending topics in the MEMBR literature

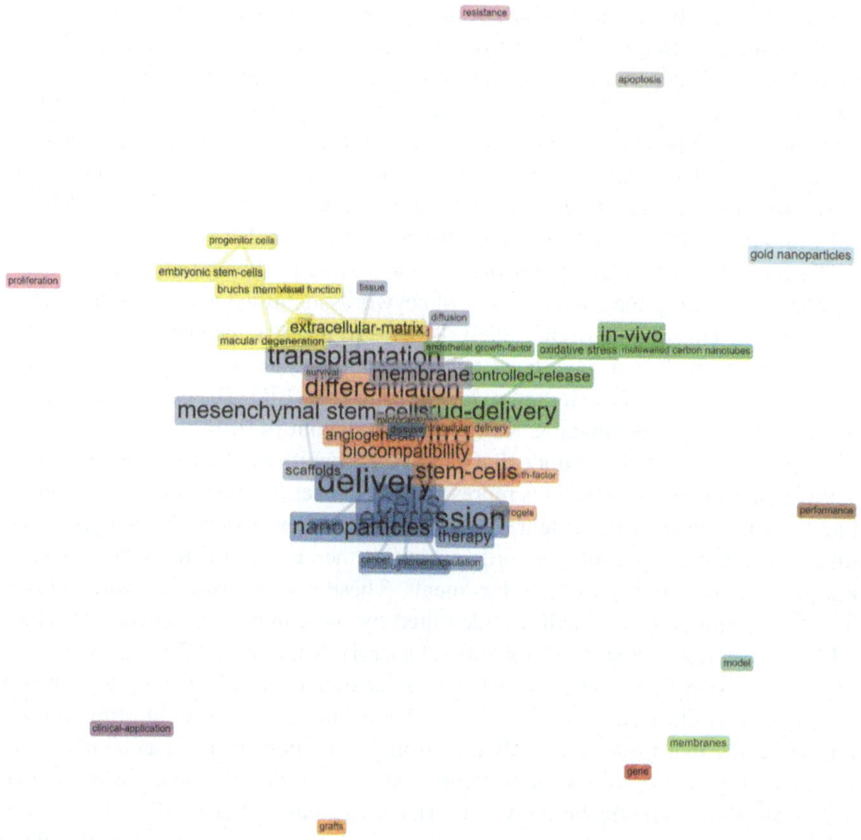

Fig. 1.10 Co-occurrence network of keywords from the Bibliometrix analysis of search MEMBR. The co-occurrence associations for those keywords are shown as a Kamada & Kawai network layout with normalization by association and using the Louvain clustering algorithm. The layout is shown for 50 nodes

Table 1.10 Co-occurrence of keywords from MEMBR listed in clusters according to the colors shown in Fig. 1.10. A keyword with a larger index of centrality (W) is positioned more centrally to the cluster. A threshold for co-occurrence clustering was set for W > 5 (keywords shaded in grey). Note that the table does not include the isolated keywords scattered at the periphery, since these keywords have indices of centrality of W = 0. In the table for Cluster 5, RPE is an acronym for retinal pigment epithelium

Cluster 1 (orange)	Cluster 2 (blue)	Cluster 3 (violet)	Cluster 4 (green)	Cluster 5 (yellow)
Fundamental biology (W)	Clinical biology (W)	Clinical biology (W)	Clinical biology (W)	Ocular biology (W)
in vitro (276.1)	cells (160.7)	transplantation (149.9)	controlled release (103.0)	Bruchs membrane (71.6)
differentiation (221.8)	expression (10.0)	membrane (125.3)	drug delivery (80.2)	RPE (64.2)
stem cells (77.1)	delivery (9.1)	survival (45.6)	oxidative stress (70.0)	extracellular matrix (40.2)
biocompatibility (59.0)		mesenchymal stem cells (13.7)	endothelial growth factor (47.8)	macular degeneration (25.9)
		scaffolds (12.9)	in vivo (36.0)	visual function (17.2)

quantify the number of times a keyword was used in each year, with the threshold for the minimum frequency set at five times in any year (Table 1.11). In the period from 2006 to 2020, there are 65 keywords that satisfy that usage restriction. Of those keywords, there is 30 identifiable as technically oriented and 33 identifiable as biologically oriented, with 2 keywords that could be in either orientation (*mechanisms*, *adsorption*, and *adhesion*). The most frequently used keywords were *nanoparticles* (37 times in 2015), *gold nanoparticles* (15 times in 2017), *delivery* (21 times in 2014), and *biomimetic synthesis* (18 times in 2013).

The most frequently used keywords for each year are shown in Fig. 1.11.

Then we quantify the co-occurrence of keywords identified in the searches shown in Table 1.1. This co-occurrence analysis helps to identify the commonality of the threads of research that have been published and is shown in Fig. 1.12 as a co-occurrence network for the literature identified in NANBIO. The co-occurrence network is shown as a Kamada & Kawai network layout [43]. In this network, each keyword is represented as a node that has a size proportional to the frequency of the keyword's occurrence. The lines represent the connections between nodes, and the line thickness reflects the extent of co-citation between nodes. The weight of the link connecting the pair of keywords quantifies the number of times that a pair of keywords co-occurs in multiple documents. These co-occurring keywords cluster into four communities, which are identified by the colors blue (cluster 1), green (cluster 2), orange (cluster 3), and pink (cluster 4). Note that the 3 isolated red keywords at the periphery were not included as separate clusters since these keywords have centrality clustering weights of W = 0. Nonetheless, the weight of the linking line indicates that these keywords are strongly co-cited in the documents. Also, there are only two keyword nodes colored purple, and despite the keyword *activation* being above the threshold (W = 12.4), the two purple keywords were not considered to form a separate cluster. We can interpret the clusters as representing three axes of research within the theme, which include *nanoparticles* (blue cluster 1, and green cluster 2), *proteins and membranes* (orange cluster 3), and *biomaterials* (pink cluster 4). The keywords at the center of the blue cluster *(nanoparticles, drug delivery)* represent the strongest co-occurring keywords in the NANBIO documents.

Table 1.12 summarizes the keyword nodes that are clustered closest to the central themes of the communities of the documents identified in the NANBIO search. There are four central themes identified as the node with the largest index of centrality, W, including *nanoparticles* and *gold nanoparticles* (nanoparticles), *protein* (proteins and membranes), and *fabrication* (biomaterials).

Conceptual Structure of BILAYER Literature

The BILAYER search was conducted to focus on lipid bilayers in nanobiotechnology using the terms *(((ALL = (lipid bilayer)) AND (ALL = (biomimet* OR bioinspir*))) AND(ALL = (nanotechnolog*)))*. These terms retrieved only 50 documents, which probably reflects the novelty of utilizing lipid bilayers in association with

Table 1.11 Frequency of occurrence of keywords for each year of publication for 345 documents of the NANBIO literature. The frequency of occurrence is indicated in the parentheses following each keyword. For inclusion in this table, we restricted the analysis to five keywords per year with a minimum word frequency of five (e.g., no keywords in 2011 matched these criteria)

Year	Keyword (frequency)	Year	Keyword (frequency)	Year	Keyword (frequency)
2020	Mesoporous silica nanoparticles (8)	2015	Nanoparticles (37)	2011	Molecular biomimetics (13)
	Doxorubicin (8)		Drug delivery (31)		Surface (10)
	Oxidative stress (5)		Nanostructures (11)		Nanotubes (7)
2019	Release (7)		Tobacco mosaic virus (6)		Polyelectrolyte multilayers (5)
	Photodynamic therapy (6)		Particles (6)	2010	peptides (9)
	Cancer (5)		Polymer (6)		DNA (7)
	Efficacy (5)	2014	Delivery (21)		Immobilization (6)
2018	Green synthesis (13)		In vitro (21)		Nanofibers (6)
	Cells (13)		Design (17)		Crystallization (5)
	System (11)		Escherichia coli (10)	2009	Self-assembled monolayers (7)
	Silver nanoparticles (11)		Quantum dots (9)		Nanotechnology (7)
	Leaf extract (9)	2013	Biomimetic synthesis (18)	2008	Surface plasmon resonance (6)
2017	Gold nanoparticles (15)		Protein (18)		Arrays (5)
	Activation (8)		Growth (14)		Display (5)
	Biosynthesis (8)		Differentiation (11)	2007	Air-water interface (12)
	Size (7)		Transport 8)		Cell-wall polymer (8)
	Nanocrystals (6)	2012	Fabrication (16)		Bacillus sphaericus CCM-2177 (7)
	Antibacterial activity (6)		Binding (13)	2006	Membranes (8)
2016	Hydrogels (10)		Proteins (12)		Ultrafiltration membranes (6)
	Adhesion (9)				
	Mesenchymal stem cells (7)		Biomaterials (10)		
	Cell (6)				
	Metal nanoparticles (6)		In vivo (9)		
	Stem cells (6)				Liposomes (5)

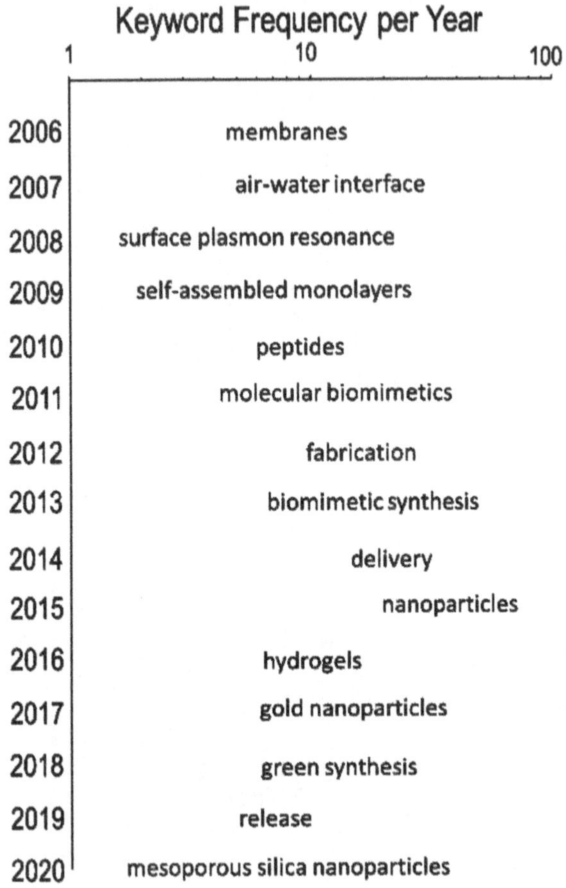

Fig. 1.11 Most frequently used keywords for each year of publication for 345 documents of the NANBIO literature. These words are the most frequently used keywords from the list shown in Table 1.11, and provide some insight into the trending topics in the NANBIO literature

nanobiotechnology (Table 1.2). We first quantify the number of times a keyword was used in each year, with the threshold for the minimum frequency set at five times in any year (Table 1.13). There were no keywords that satisfied that criterion for the years 2012, 2013, 2015, 2019, 2020, or 2021. In the period from 2011 to 2018, there are five keywords that satisfy that usage restriction. Of those keywords, there are two identifiable as technically oriented and three identifiable as biologically oriented.

These keywords for each year are shown in Fig. 1.13.

Then we quantify the co-occurrence of keywords identified in the searches shown in Table 1.1. This co-occurrence analysis helps to identify the commonality of the threads of research that have been published and is shown in Fig. 1.14 as a co-occurrence network for the literature identified in BILAYER. The co-occurrence network is shown as a Kamada & Kawai network layout [43]. In this network, each keyword is represented as a node that has a size proportional to the frequency of the keyword's occurrence. The lines represent the connections between nodes, and the

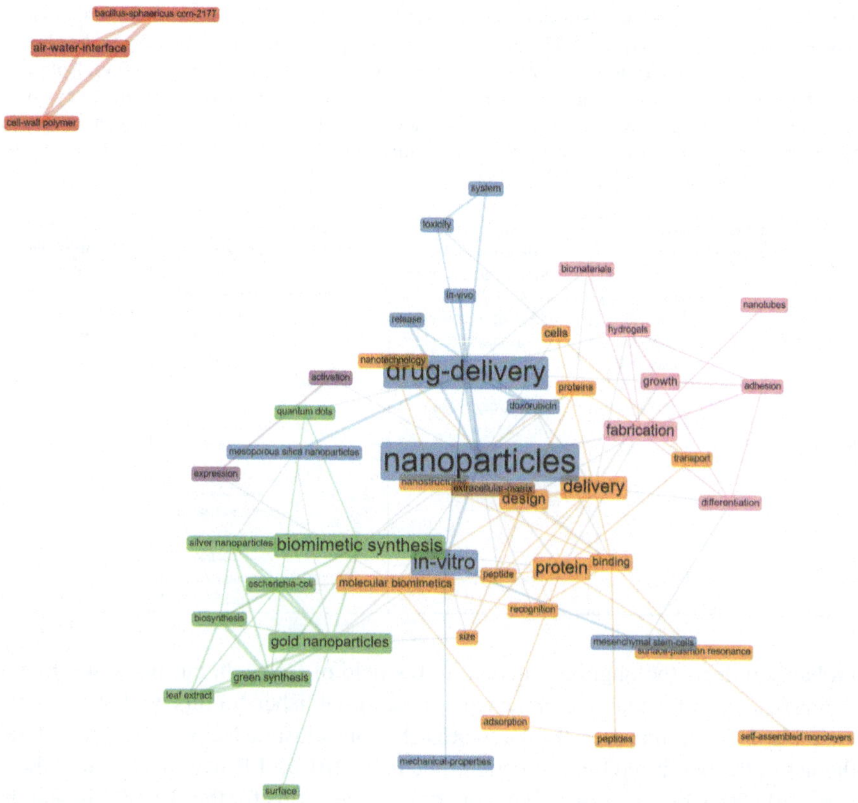

Fig. 1.12 Co-occurrence network of keywords from the Bibliometrix analysis of search NANBIO. The co-occurrence associations for those keywords are shown as a Kamada & Kawai network layout with normalization by association and using the Louvain clustering algorithm. The layout is shown for 50 nodes. Although the red nodes are linked strongly, the index of centrality is W = 0, and these are not included in Table 1.12

line thickness reflects the extent of co-citation between nodes. The weight of the link connecting the pair of keywords quantifies the number of times that a pair of keywords co-occurs in multiple documents. Due to the small number of documents, these co-occurring keywords cluster only into two communities, which are identified by the colors violet (cluster 1) and green (cluster 2). It is difficult to identify those communities with separate themes, so we can adopt the themes utilized in the conceptual structure of the other searches to label the clusters as representing the themes *proteins and membranes* (violet cluster 1) *and clinical biology* (green cluster 2). Note that many of the other keyword nodes include reasonably strong links between a small number of nodes, with the majority of keywords mapped in the network as single nodes. The majority of keywords had indices of centrality of W = 0 since they were distributed widely around the network. This wide distribution

Table 1.12 Co-occurrence of keywords from NANBIO listed in clusters according to the colors shown in Fig. 1.12. A keyword with a larger weight (W) is positioned more centrally to the cluster. A threshold for co-occurrence clustering was set for W > 5 (keywords shaded in grey). Note that the table does not include the three isolated red keywords at the periphery, since these keywords have centrality clustering weights of W = 0. There are only two keyword nodes colored purple, and despite the keyword *activation* being above the threshold (W = 12.4), the two purple keywords were not considered to form a separate cluster

Cluster 1 (blue)	Cluster 2 (green)	Cluster 3 (orange)	Cluster 4 (pink)
Nanoparticles (W)	Nanoparticles (W)	Proteins and membranes (W)	Biomaterials (W)
nanoparticles (325.9)	gold nanoparticles (118.6)	protein (104.2)	fabrication (94.6)
drug delivery (198.5)	biomimetic synthesis (116.4)	molecular biomimetics (88.4)	growth (14.2)
in vitro (99.5)	Escherichia coli (18.5)	binding (76.7)	differentiation (11.4)
mesoporous silica nanoparticles (25.5)	quantum dots (11.2)	delivery (72.8)	hydrogels (7.1)
doxorubicin (8.2)	green synthesis (6.3)	design (56.6)	biomaterials (3.0)
mesenchymal stem cells (5.1)	silver nanoparticles (4.2)	surface plasmon resonance (44.0)	adhesion (1.5)
toxicity (2.9)	biosynthesis (1.8)	cells (15.4)	
		nanostructures (15.3)	
		proteins (14.3)	
		size (14.0)	
		adsorption (11.0)	
		nanotechnology (5.0)	
		transport (3.1)	
		peptide (3.0)	
		recognition (2.3)	
		peptides (1.1)	

probably reflects the ongoing interest in this field of research, but without a major thematic axis within nanomedicine having been established for this field of research.

Table 1.14 summarizes the keywords that are clustered closest to the central themes of the published literature identified in the BILAYER search. Note that there were only 50 published documents found using the terms for the BILAYER search, which were too few to allow strong clustering of keywords to be found. Nonetheless, the themes of *liposomes*, *membrane proteins*, and *drug delivery* did emerge in the conceptual structure of this field.

Conceptual Structure of NANSENS Literature

The NANSENS search was conducted to focus on bioinspired nanosensors using the terms *((ALL = (biomimet* OR bioinspir*)) AND (ALL = (nanosens*)))*. These terms retrieved 118 documents (Table 1.2). We first quantify the number of times a keyword was used in each year, with the threshold for the minimum frequency set at five times in any year (Table 1.15). In the period from 2008 to 2020, there are 18 keywords that satisfy that usage restriction. Of those keywords, there are 14 identifiable as technically oriented and 4 keywords that could also be biologically oriented (*selective detection*, *recognition*, *sensitivity*, and *design*). The most frequently used keywords were *nanoparticles* (14 times in 2019) and *actuators* (10 times in 2008) (Table 1.15).

Table 1.13 Frequency of occurrence of keywords for each year of publication for 50 documents of the BILAYER literature. The frequency of occurrence is indicated in the parentheses following each keyword. For inclusion in this table, we restricted the analysis to five keywords per year with a minimum word frequency of five (e.g., no keywords in 2012, 2013, 2015, 2019, 2020, or 2021 matched these criteria)

Year	Keyword (frequency)
2018	nanoparticles (5)
2017	Drug delivery (5)
2016	Binding (6)
2014	Liposomes (6)
2011	Lipid bilayers (6)

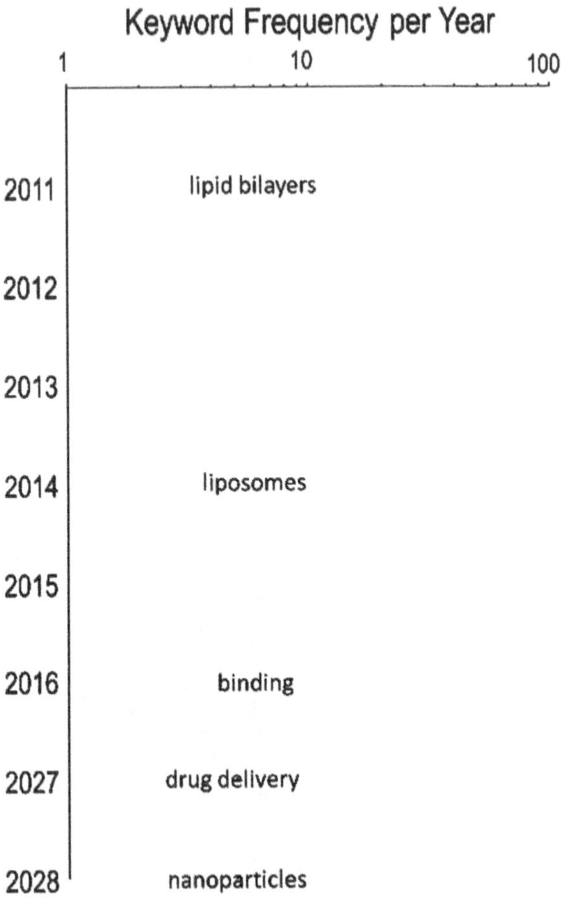

Fig. 1.13 Most frequently used keywords for each year of publication for 50 documents of the BILAYER literature. These words are the most frequently used keywords from the list shown in Table 1.13 and provide some insight into the trending topics in the BILAYER literature. Note that the number of documents was not sufficient to identify keywords with a minimum frequency for the years of 2012, 2013, and 2015

Then we quantify the co-occurrence of keywords identified in the searches shown in Table 1.1. This co-occurrence analysis helps to identify the commonality of the threads of research that have been published and is shown in Fig. 1.16 as a co-occurrence network for the literature identified in BILAYER. The co-occurrence network is shown as a Kamada & Kawai network layout [43]. In this network, each keyword is represented as a node that has a size proportional to the frequency of the keyword's occurrence. The lines represent the connections between nodes, and the line thickness reflects the extent of co-citation between nodes. The weight of the link connecting the pair of keywords quantifies the number of times that a pair of keywords co-occurs in multiple documents. These co-occurring keywords cluster only into four communities, which are identified by the colors grey (cluster 1), blue (cluster 2), green (cluster 3), and teal (cluster 4). Note that many of the other keyword nodes include links between a small number of nodes, with the majority of keywords mapped in the network as single nodes. Those isolated keywords had indices of centrality of $W = 0$ since they were distributed widely around the

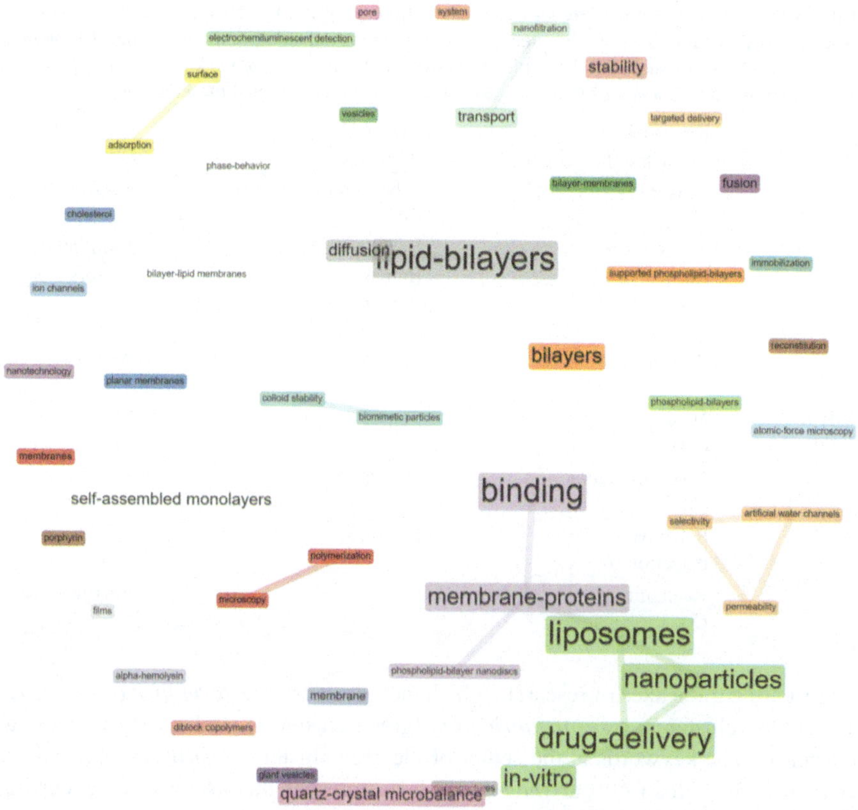

Fig. 1.14 Co-occurrence network of keywords from the Bibliometrix analysis of search BILAYER. The co-occurrence associations for those keywords are shown as a Kamada & Kawai network layout with normalisation by association and using the Louvain clustering algorithm. The layout is shown for 50 nodes

Table 1.14 Co-occurrence of keywords from BILAYER listed in clusters according to the colors shown in Fig. 1.14. A keyword with a larger weight (W) is positioned more centrally to the cluster. Note that there were only 50 published documents, which were insufficient numbers of documents to establish strong clustering of the keywords. A threshold for co-occurrence clustering was set for W > 5 (keywords shaded in grey). Note that the table does not include the isolated keywords scattered at the periphery, since these keywords have centrality clustering weights of W = 0

Cluster 1 (violet)	Cluster 2 (green)
Proteins and membranes (W)	**Clinical biology (W)**
membrane proteins (9.0)	liposomes (9.0)
	drug delivery (5.0)

network. This wide distribution probably reflects the ongoing interest in this field of research, but without additional major thematic axes within nanomedicine having been established for this field of research. We can interpret the four clusters as

Table 1.15 Frequency of occurrence of keywords for each year of publication for 118 documents of the NANSEN literature. The frequency of occurrence is indicated in the parentheses following each keyword. For inclusion in this table, we restricted the analysis to five keywords per year with a minimum word frequency of five (e.g., no keywords in 2011 matched these criteria)

Year	Keyword (frequency)	Year	Keyword (frequency)	Year	Keyword (frequency)
2020	Sensor (6)	2018	Fabrication (8)	2013	Sensors (9)
	Nanosensor (6)		Enhanced Raman spectroscopy (5)		Biomimetic sensors (5)
	Biosensor (5)		Spectroscopy (5)	2008	Actuators (10)
2019	Nanoparticles (14)	2017	Sensitivity (9)		Membrane (7)
	Quantum dots (6)		Design (5)		
	Selective detection (6)		Films (5)		
	Recognition (5)				Polymers (5)

representing three axes of research, which include *sensor performance* (grey cluster 1 and blue cluster 2), *sensor fabrication* (green cluster 3), and *spectroscopy* (teal cluster 4). The keywords at the center of the grey cluster *(sensitivity)*, blue cluster (*sensors*), and green cluster (*nanoparticles*) represent the strongest co-occurring keywords in the NANSENS documents, with W > 150.

Table 1.16 summarizes the keywords that are clustered closest to the central themes of the published literature identified in the NANSEN search. Those central themes include *sensors*, *sensitivity*, *nanoparticles*, and *spectroscopy*. The other main keywords can be interpreted as being related to the method and application of those technologies.

Summary of the Conceptual Structure of the Nanomedicine Literature: Technological

We can consider that the conceptual structure of the technological nanomedicine literature is clustered around four central themes of research, which are *fundamental biology*, *clinical biology*, *nanoparticles*, and *sensors (performance, fabrication)*. Within those clustering themes, the dominant keyword nodes were identified as those with indices of centrality greater than 150. These dominant keyword nodes are in vitro, *differentiation*, *cells*, *transplantation*, *nanoparticles*, *drug delivery*, *sensors*, and *sensitivity.*

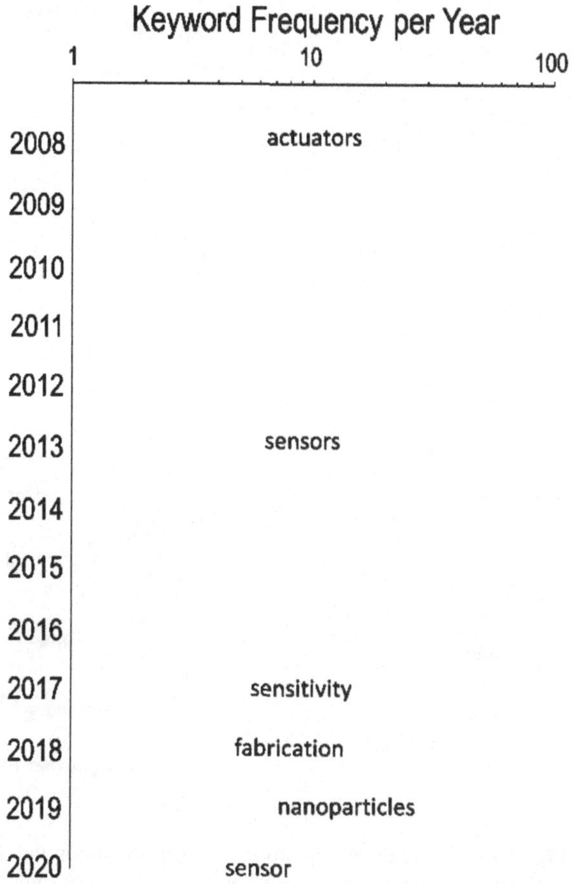

Fig. 1.15 Most frequently used keywords for each year of publication for 118 documents of the NANSEN literature. These words are the most frequently used keywords from the list shown in Table 14 and provide some insight into the trending topics in the NANSEN literature

Reflections on the Conceptual Structure of Nanomedicine and Emerging Nanobiotechnologies

The field of nanomedicine is large, particularly when trying to identify nanobiotechnologies that are pertinent to the state-of-the-art and also to the future directions in the field of nanomedicine. The central themes to date can be identified from the conceptual structure that was identified from the bibliometrix analysis of the literature searches (Table 1.17).

The striking feature of that conceptual structure is the dominant role of nanoparticles across several of the central themes of nanomedicine research. That dominance is not surprising, given that nanoparticles are the most recognizable aspect of nanotechnology. Nanoparticles currently provide the most commonly used nanostructured systems for drug delivery, diagnostics, and sensors.

We can also use the bibliometrix analysis to identify emerging technologies that could shape and consolidate the future conceptual structure of nanomedicine. We

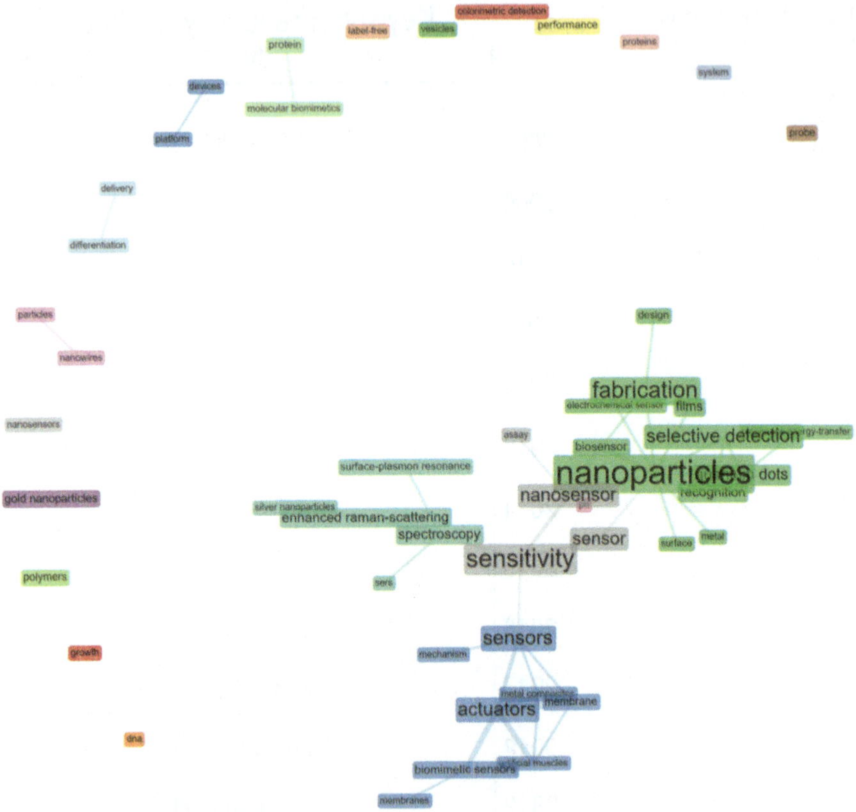

Fig. 1.16 Co-occurrence network of keywords from the Bibliometrix analysis of search NANSEN. The co-occurrence associations for those keywords are shown as a Kamada & Kawai network layout with normalization by association and using the Louvain clustering algorithm. The layout is shown for 50 nodes

Table 1.16 Co-occurrence of keywords from NANSEN listed in clusters according to the colors shown in Fig. 1.16. A keyword with a larger weight (W) is positioned more centrally to the cluster. A threshold for co-occurrence clustering was set for W > 5 (keywords shaded in grey). Note that the table does not include the isolated keywords scattered at the periphery, since these keywords have centrality clustering weights of W = 0. In the table for Cluster 14, RPE is an acronym for retinal pigment epithelium

Cluster 1 (grey)	Cluster 2 (blue)	Cluster 3 (green)	Cluster 4 (teal)
Sensor performance (W)	Sensor performance (W)	Sensor fabrication (W)	Spectroscopy (W)
sensitivity (236.2)	sensors (153.0)	nanoparticles (205.8)	spectroscopy (101.0)
nanosensor (137.6)	actuators (62.8)	fabrication (27.0)	enhanced Raman spectroscopy (27.0)
sensor (57.4)	biomimetic sensors (27.0)	selective detection (14.0)	
		biosensor (12.5)	
		quantum dots (12.5)	

Table 1.17 Conceptual structure of the nanomedicine literature identified from all the Bibliometrix analyses of the searches shown in Table 1.1

Central Theme	Central keyword
Fundamental biology	Drug delivery
	In vitro
	Differentiation
Clinical biology	Nanoparticles
	Cells
	Transplantation
Microbiota	Gut microbiota
Nanoparticles	Nanoparticles
	Drug delivery
Sensors (performance, fabrication)	Sensitivity
	Sensors
	Nanoparticles

consider an emerging technology as one that is not already clustered centrally in the nanomedicine literature but is gaining momentum in the keyword conceptual structure of research and publications in nanomedicine. Thus, we can start by identifying the keywords with indices of centrality of $W < 5$ from Tables 1.4, 1.6, 1.8, and 1.12. These keywords are listed in Table 1.18. Note that keywords with an index of centrality of $W = 0$ were not included in Table 1.18. The keywords are linked to the themes that were identified from the bibliometrix analyses.

The lists of non-central keywords and themes provide a rational basis to identify an emerging conceptual structure for nanomedicine, which is illustrated in Fig. 1.17. For clarity, this figure includes only the overlapping keywords that appear in multiple themes. Indeed, these overlapping keywords provide pointers to the evolution of the conceptual structure, particularly if we consider that *biomedical membrane technology* is synonymous with theme *proteins and membranes*. The emergence of *biological engineering* as a central theme would then supersede *proteins and membranes* for the growth of the field of nanomedicine.

The emergence of biological engineering includes technologies that are oriented toward bioinspired and biomimetic technologies. One could consider that such technologies have a hierarchy that includes the three levels of *(i)* targeting the genome with DNA, *(ii)* targeting the membrane with proteins, and *(iii)* targeting the microbiota with bioinspired devices. The topic of the microbiota is an exploding field where there is scope for introducing such new technologies for diagnosis and therapy. It is clear that the emergence of biological engineering as a major theme, or indeed as a discipline, relies upon the combination of expertise from the traditional disciplines of chemistry, physics, biology, engineering, and mathematics. The significant feature for the growth of nanomedicine and the emergence of biological engineering is that these require a multidisciplinary approach and are not to be driven only as a collection of expertise from the traditional disciplines.

Managing the emergence of biological engineering to optimize its multidisciplinary contribution to outcomes in nanomedicine requires a non-traditional approach. Otherwise, the growth of nanomedicine will remain constrained by the dichotomy of whether the starting point for advances in nanobiotechnology and nanomedicine is in the realm of fundamental scientists in traditional disciplines to first seek knowledge in the basic functions of molecules that comprise biomimetic membranes or in the realm of clinicians and engineers who first seek to observe and characterize defined medical problems and needs. We can resolve this dichotomy by adopting the approach of biological engineering, which results in multidisciplinary-targeted outcomes from research that requires a team with interdisciplinary expertise. A biological engineering approach, which includes biomimetic membrane technology, commences with the premise that the value of nanomedicine is to improve the health of individuals. A related premise is that the value of biomimetic membrane technology provides a symbiotic interface to improve the function of systems and devices that interact with tissues and cells of the body [2]. Indeed, such a biological engineering approach allows contributions from both a fundamental and an applied approach within the same project. This is because the multidisciplinary outcomes are the primary drivers of such a biological engineering approach.

Table 1.18 Non-central keywords in the conceptual structure of the nanomedicine literature. These keywords were identified by indices of centrality of W < 5. Keywords with W = 0 are not included in these lists

Theme	Keywords with W < 5	Theme	Keywords with W < 5	Theme	Keywords with W < 5
Fundamental biology	Protein	Clinical biology	Cell	Therapeutic applications	Iron-oxide nanoparticles
	Peptide		Tumor		Biomedical applications
	Scaffolds		Growth		Cancer therapy
	Extracellular matrix		Liposomes		Targeted delivery
	Design		Vesicles		Quantum dots
	Fabrication		Membrane		Magnetic nanoparticles
	Biomedical applications		Bone		Biomimetic nanoparticles
	Chitosan		Binding		Functionalisation
	Blood brain barrier		Mechanisms		Coated nanoparticles
	Tissue		Design		Platform
	Cancer therapy		Model		Mesoporous silica nanoparticles
	Differentiation		Expression		Biomimetic nanoparticles
	Stem cells		Release		Coated nanoparticles
	Gene delivery		Drug		Magnetic nanoparticles
	Controlled release		Resistance		Dendritic cells
	Targeted delivery		Doxorubicin	Microbiota	Impact
	Iron oxide nanoparticles		Paclitaxel		Colitis
	Gold nanoparticles		Nanocarriers		Gene expression
	Quantum dots		Nanotechnology		Association
Nanoparticles	Toxicity		Nanomaterials		Gut
	Silver nanoparticles	Bacterial	Obesity		Risk
	Biosynthesis		Evolution		Infection
Proteins and membranes	Transport		Sequences		Children
	Peptide		Colonisation		Mechanisms
	Recognition		Microbiome		Acid
	Peptides		Human gut microbiome		Double-blind
Biomaterials	Biomaterials		Reveals		Gene
	Adhesion	Microbiology	Mice		Resistance
			Activation		
			Ulcerative colitis		Prevalences
			In vitro		Cardiovascular disease

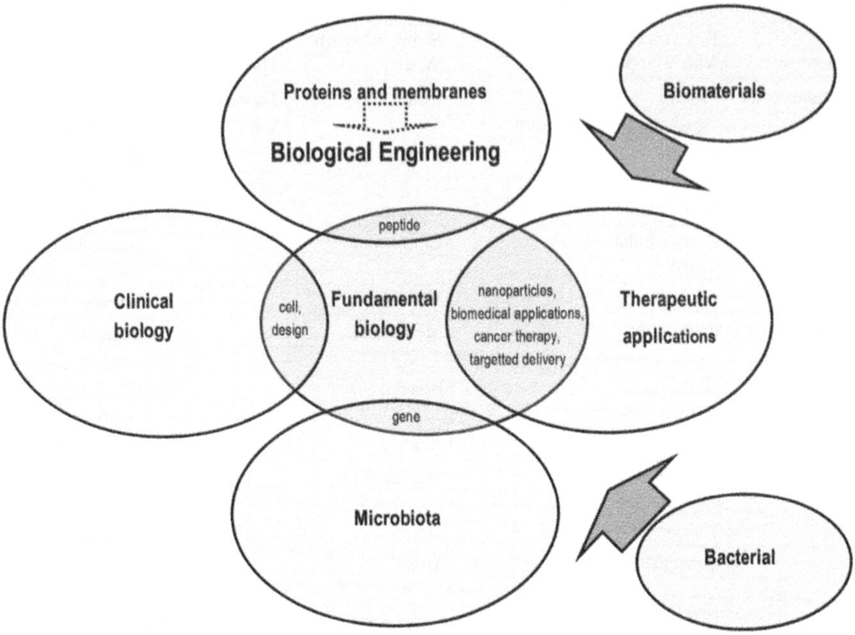

Fig. 1.17 Non-central keywords and associated themes in the conceptual structure of the nanomedicine literature. These keywords were identified by indices of centrality of W < 5. Keywords with W = 0 are not included in these lists. The major thematic axis of *Biological Engineering* in nanomedicine is most likely to emerge and to supersede the theme of *proteins and membranes*, as indicated by the dashed arrow in the uppermost ellipse. The themes of *biomaterials* and *bacterial* are most likely to contribute to the emergence of other themes that will consolidate the overall conceptual structure of the nanomedicine field

This in turn drives the conceptual structure of the knowledge that is required to be able to best biologically engineer a nanobiotechnological solution. Indeed, these nanobiotechnological advances support the advancement of nanomedicine.

In the following chapters, we will discuss ways to manage outcomes from a multidisciplinary approach in biological engineering, nanobiotechnologies, and nanomedicine. We will also discuss the impact and perspectives on medical diagnostics and therapeutics from two aspects. The first aspect is to identify diseases that are attributable to a dysfunction in the cell membrane and/or the incorporated membrane transport proteins. That description also discusses the types of diseases that can be amenable to detailed study using in vitro biomimetic membrane technologies. Although it is clear that understanding the consequences and syndrome of diseases depends on studies in the complete organism, the second aspect is to identify diseases where there can be an advantage in utilizing in vitro biomimetic membrane systems to investigate cause-and-effect relationships at a more fundamental level. For example, a greater use of in vitro biomimetic membrane systems may provide insights into therapeutic molecules that may provide more directed therapeutic approaches for the entire organism, which may also help to minimize side effects of therapeutic medications.

The significant feature of biological engineering, and particularly biomimetic membrane technology, is that these are emerging disciplines that introduce a paradigm shift in the way that nanobiotechnologies are used in nanomedicine. The paradigm shift of this new field of biological engineering is that a nanostructured biomimetic membrane is used in medical devices to provide an interface that is self-assembled in similar ways to the cells of the body. This provides the medical device with an interface that will be capable of communicating biological molecules with the body. The medical device will have the capability to function autonomously to assist the functions of cells and tissues in the body. This paradigm shift implicates that the framing of the problem is the driver for the multidisciplinary research needed to provide a solution, utilizing interdisciplinary skills that are embodied in the emerging field of biological engineering.

References

1. Martin, D. (2007). *Nanobiotechnology of biomimetic membranes*. Springer.
2. Alcaraz, J. P., Cinquin, P., & Martin, D. K. (2018). Tackling the concept of symbiotic implantable medical devices with nanobiotechnologies. *Biotechnology Journal, 13*, 1800102.
3. Robertson, J., Yealland, G., Avila-Olias, M., Chierico, L., Brandmann, O., Renshaw, S., & Battaglia, G. (2014). pH-sensitive tubular polymerosomes: Function and applications in cellular delivery. *ACS Nano, 8*, 4650–4661.
4. Blanazs, A., Armes, S., & Ryan, A. (2009). Self-assembled block copolymer aggregates: From micelles to vesicles and their biological applications. *Macromolecular Rapid Communications, 30*, 267–277.
5. Blodgett, K. (1935). Films built by depositing successive monomolecular layers on a solid surface. *Journal of the American Chemical Society, 57*, 1007–1022.
6. Blodgett, K., & Langmuir, I. (1937). Built-up films of barium stearate and their optical properties. *Physical Review, 51*, 964–982.
7. Blodgett, K. (1939). A method of extinguishing the reflection of light from glass. *Science, 89*, 60–61.
8. Johonnott, E. (1899). XLVIII thickness of the black spot in liquid films. *Philosophical Magazine Series, 5*(47), 510–522.
9. Johonnott, E. (1906). LXVIII The Black spot in thin liquid films. *Philosophical Magazine Series, 6*(11), 746–753.
10. Folch, J., & Lees, M. (1951). Proteolipids, a new type of tissue lipoproteins – their isolation from brain. *Journal of Biological Chemistry, 191*, 807–817.
11. Mueller, P., Rudin, D., Ti Tien, H., & Wescott, W. (1962). Reconstitution of cell membrane structure in vitro. *Circulation, 26*, 1167–1171.
12. Mueller, P., Rudin, D., Ti Tien, H., & Wescott, W. (1962). Reconstitution of cell membrane structure in vitro and its transformation into an excitable system. *Nature, 194*, 979–980.
13. Bangham, A., Standish, M., & Watkins, J. (1965). Diffusion of univalent ions across the lamellae of swollen phospholipids. *Journal of Molecular Biology, 13*, 238–252.
14. Cornell, B., Braach-Maksvytis, V., King, L., Osman, P., Raguse, B., Wieczorek, L., & Pace, R. (1997). A biosensor that uses ion-channel switches. *Nature, 387*, 580–583.
15. Lang, H., Duschl, C., & Vogel, H. (1994). A new class of thiolipids for the attachment of lipid bilayers on gold surfaces. *Langmuir, 10*, 197–210.

16. Naumann, R., Jonczyk, A., Kopp, R., van Esch, J., Ringsdorf, H., & Knoll, W. G. (1995). Incorporation of membrane proteins in solid-supported lipid bilayers. *Angewandte Chemie International Edition, 34*, 2056–2058.
17. Schmidt, E. K., Liebermann, T., Kreiter, M., Jonczyk, A., Naumann, R., Offenhäusser, A., Neumann, E., Kukol, A., Maelicke, A., & Knoll, W. (1998). Incorporation of the acetylcholine receptor dimer from Torpedo californica in a peptide supported lipid membrane investigated by surface plasmon and fluorescence spectroscopy. *Biosensors Bioelectronics, 13*, 585–591.
18. Soranzo, T., Martin, D. K., Lenormand, J., & Watkins, E. (2017). Coupling neutron reflectivity with cell-free protein synthesis to probe membrane protein structure in supported bilayers. *Scientific Reports, 7*, 3399.
19. Ma, W. (2016). The essence of life. *Biology Direct, 11*, 49.
20. Ciamician, G. (1912). The photochemistry of the future. *Science, 36*, 385–394.
21. Majumdar, P., Pant, D., & Patra, S. (2017). Integrated photobioelectrochemical systems: A paradigm shift in artificial photosynthesis. *Trends in Biotechnology, 35*, 285–287.
22. Bricarello, D., Smilowitz, J., Zivkovic, A., German, J., & Parikh, A. (2011). Reconstituted lipoprotein: A versatile class of biologically-inspired nanostructures. *ACS Nano, 5*, 42–57.
23. Haynie, D., Zhang, L., Zhao, W., & Rudra, J. (2006). Protein-inspired multilayer nanofilms: Science, technology and medicine. *Nanomedicine, 2*, 150–157.
24. Iler, R. (1966). Multilayers of colloidal particles. *Journal of Colloid and Interface Science, 21*, 569–594.
25. Decher, G. (1997). Fuzzy nanoassemblies: Toward layered polymeric multicomposites. *Science, 277*, 1232–1237.
26. Martin, C., & Kohli, P. (2003). The emerging field of nanotube biotechnology. *Nature Reviews Drug Discovery, 2*, 29–37.
27. Peters, R. (2005). The nanopore connection to cell membrane unitary permeability. *Traffic, 6*, 199–204.
28. Garni, M., Thamboo, S., Schoenenberger, C., & Palivan, C. (2017). Biopores/membrane proteins in synthetic polymer membranes. *Biochimica et Biophysica Acta, 1859*, 619–638.
29. Banerji, S., & Hayes, M. (2007). Examination of nonendocytic bulk transport of nanoparticles across phospholipid membranes. *Langmuir, 23*, 3305–3313.
30. He, B., Yuan, L., Dai, W., Zhang, H., Wang, X., Fang, W., & Zhang, Q. (2016). Dynamic bioadhesion of polymer nanoparticles on MDCK epithelial cells and its impact on bio-membranes, endocytosis and paracytosis. *Nanoscale, 8*, 6129–6145.
31. Plebanek, M., Mutharasan, R., Volpert, O., Matov, A., Gatlin, J., & Thaxton, C. (2015). Nanoparticle targeting and cholesterol flux throgh scavenger receptor type B-1 inhibits cellular exosome uptake. *Scientific Reports, 5*, 15724.
32. Kooijmans, S. A. A., Vader, P., van Dommelen, S. M., van Solinge, W. W., & Schiffelers, R. M. (2012). Exosome mimetics: A novelclass of drug delivery systems. *International Journal of Nanomedicine, 4*, 1525–1541.
33. Tanner, P., Baumann, P., Enea, R., Onaca, O., Palivan, C., & Meier, W. (2011). Polymeric vescicles: From drug carriers to nanoreactors and artificial organelles. *Accounts of Chemical Research, 44*, 1039–1049.
34. De Oliveira, H., Thevenot, J., & Lecommandoux, S. (2012). Smart polymerosomes for therapy and diagnosis: Fast progress toward multifunctional biomimetic nanomedicines. *Wiley Interdisciplinary Reviews Nanomedicine and Nanobiotechnology, 4*, 525–546.
35. Pierstorff, E., & Ho, D. (2007). Monitoring, diagnostic, and therapeutic technologies for nanoscale medicine. *Journal of Nanoscience and Nanotechnology, 7*, 2949–2968.
36. Wang, L., Tang, K., Chiu, S., Yang, S., & Kuo, C. (2011). A bio-inspired two-layer multiple-walled carbon nanotube–polymer composite sensor array and a bio-inspired fast-adaptive readout circuit for a portable electronic nose. *Biosensors Bioelectronics, 26*, 4301–4307.
37. Siontorou, C., & Batzias, F. (2014). A methodological combined framework for roadmapping biosensor research: A fault tree analysis approach within a strategic technology evaluattion frame. *Critical Reviews in Biotechnology, 34*, 31–55.

38. European Science Foundation. (2005, January 15). *ESF Forward Look report on Nanomedicine*. (www.esf.org).
39. Freitas, R. A. (2005). What is nanomedicine ? *Nanomedicine, 1*, 2–9.
40. Duncan, R. (2005). Nanomedicine gets clinical. *Materials Today, 8*, 16–17.
41. Heath, J. R. (2015). Nanotechnologies for biomedical science and translational medicine. *PNAS, 112*, 14436–14443.
42. Aria, M., & Cuccurullo, C. (2017). Bibliometrix: An R-tool for comprehensive science mapping analysis. *Journal of Informetrics, 11*, 929–975.
43. Kamada, T., & Kawai, S. (1989). An algorithm for drawing general undriected graphs. *Information Processing Letters, 31*, 7–15.

Chapter 2
Nanostructured Systems and the Value of Biological Engineering

This chapter explores the design and assembly of nanostructured systems with inspiration from biology. Here we consider the biological inspiration as not the approach typically adopted, which is to construct synthetic materials inspired from nature. In contrast, the approach described in this chapter utilizes biological components in the assembly of smart nanostructured systems. We adapt the term "biological engineering" to describe a discipline that embodies this approach. Indeed, in 1970, this term was introduced formally with the intention to integrate engineering with biological systems to move beyond single disciplinary areas such as medicine, agriculture, or fermentation engineering. In this chapter, we further refine the discipline of "biological engineering" to be one that utilizes biological proteins, molecules, and lipids in combination with synthetic materials to assemble smart nanostructural systems. We illustrate this discipline with the assembly of nanostructured systems that are targeted for applications such as diagnostic, therapeutic, biofuel cell, or tissue enhancement/replacement in the body.

Nanostructured Systems

We must first describe what we mean by a nanostructured system. Many devices incorporate components that are nanostructured, whether by default or by design. However, those devices are usually large enough to be observed, manipulated, and capable of providing measurements that can be interfaced. All of those characteristics would not necessarily define such a system as being truly nanostructured. We can begin by collating the thoughts from the scientific literature on the constitution of a nanostructured system. The WordCloud of Fig. 2.1 illustrates the 50 most frequently used terms related to nanostructures systems. These words were found from a search of publications in the Clarivate® Web of Science (WoS) using the terms *ALL = (biological engineering) AND (ALL = (smart systems)) AND ALL = (nanostruct*).*

D. Martin, I. Vilgrain, *Biomimetic Membrane Technology*,
https://doi.org/10.1007/978-3-031-90594-0_2

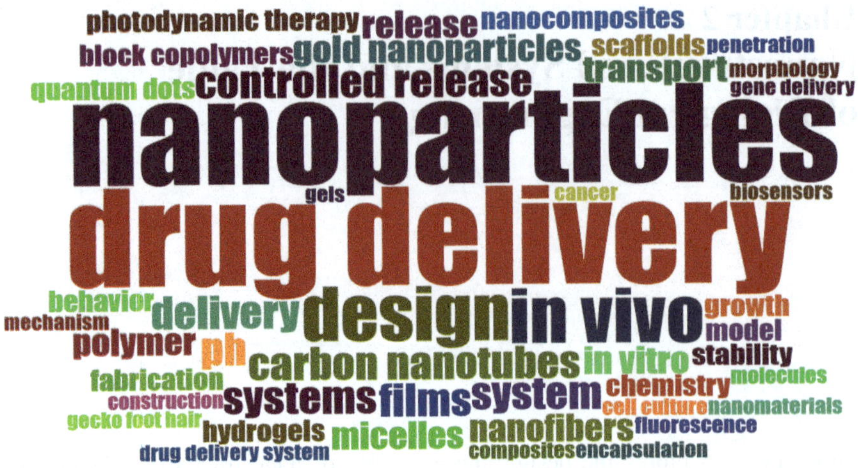

Fig. 2.1 WordCloud of the 50 most frequent words from a bibliometric analysis of the literature in the WoS using the search terms *ALL = (biological engineering) AND (ALL = (smart systems)) AND ALL = (nanostructur*)*. The literature search retrieved 121 items. The analysis was performed using the "Bibliometrix" library of RStudio v1.2.5033 [1]

This WordCloud does not identify a main unifying theme, most likely because nanostructure refers to a length scale. This does not place it in a specific domain of a particular scientific or engineering discipline. Nonetheless, we can gain further insight into defining smart nanostructured systems that utilize biological engineering by looking in more detail at the publications with more than 100 citations. These publications can be considered as being the most influential in the field and are listed in Table 2.1. Those highly cited articles were published in the 14-year period from 2004 to 2018. We can also consider the current research directions that influence the authors in this field by looking to the articles that have been published in 2022, which are listed in Table 2.2.

To move further toward identifying a main unifying theme to define smart nanostructured systems that utilize biological engineering, we can compare the main WoS disciplines for those highly cited publications (Table 2.3) with the main WoS disciplines of the recent 2022 publications (Table 2.4).

Table 2.3 suggests that in the period 2004 to 2028, a nanostructured system was categorized mainly on a discipline-basis, and particularly as a system that comprised nanoparticles. The traditional engineering disciplines dominated the categories of highly cited publications from this period. Table 2.4 suggests that the publications of 2022 are less categorized by the more traditional disciplines in engineering, chemistry, and physics. The 2022 publications are included in more biotechnology and applied chemistry disciplines. Such a distribution broadens the publication of nanostructured systems into applied biological, chemical, and polymer categories.

This bibliometric analysis provides some pointers to what we can consider as a nanostructured system. The literature is moving away from the concept that

Table 2.1 The top 14 publications with more than 100 citations from the WoS databases. These publications were identified from a literature search of Wos using the search terms *ALL = (biological engineering) AND (ALL = (smart systems)) AND ALL = (nanostructur*)*. The literature search retrieved 121 items

Article title	Authors	Reference	Citations
Gold nanocages covered by smart polymers for controlled release with near-infrared light	Yavuz MS, Cheng Y, Chen J, Cobley CM, Zhang Q, Rycenga M, Xie J, Kim C, Song KH, Schwartz AG, Wang LV, Xia Y	*Nat. Mater.*, 8:935–939, 2009	1179
Bio-inspired design of multiscale structures for function integration	Liu K, Jiang L	*NANO Today*, 6:155–175, 2011	567
A review of atomic force microscopy imaging systems: application to molecular metrology and biological devices	Jalili N, Laxminarayana K	*Mechatronics*, 14:907–945, 2004	394
Smart nanoparticles for drug delivery application: development of versatile nanocarrier platforms in biotechnology and nanomedicine	Lombardo D, Kiselev MA, Caccamo MT	*J. Nanomater.*, 2019:3702518, 2019	385
DNA nanotechnology based on i-motif structures	Dong Y, Yang Z, Liu D	*Acc. Chem. Res.*, 47:1853–1860, 2014	257
Engineering nanoscale order into a designed protein fiber	Papapostolou D, Smith AM, Atkins EDT, Oliver SJ, Ryandnov MG, Serpell LC, Woolfson DN	*PNAS*, 104:10853–10858, 2007	200
Silica coated upconversion nanoparticles: a versatile platform for the development of efficient theranostics	Liu JN, Bu WB, Shi JL	*Acc. Chem. Res.*, 48:1797–1805, 2015	189
Dynamic supramolecular complexes constructed by orthogonal self-assembly	Hu XY, Xiao T, Lin C, Huang F, Wang L	*Acc. Chem. Res.*, 47:2941–2051, 2014	189
Virus-directed design of a flexible BaTiO3 nanogenerator	Jeong CK, Kim I, Park KI, Oh H, Paik H, Hwang GT, No K, Nam YS, Lee KJ	*ACS Nano*, 7:11016–11025, 2013	179
Methods of synthesis, properties and biomedical applications of CuO nanoparticles	Grigore ME, Biscu ER, Holban M, Gestal MC, Grumezescu AM	*Pharmaceuticals*, 9:75, 2016	172

(continued)

Table 2.1 (continued)

Article title	Authors	Reference	Citations
Hierarchical micro and nano structured, hydrophilic, superhydrophobic and superoleophobic surfaces incorporated in microfluidics, microarrays and lab on chip microsystems	Gogolides E, Ellinas K, Tserepi A	*Microelectr. Eng.*, 132:135–155, 2015	155
RGD-peptide functionalized graphene biomimetic live-cell sensor for real-time detection of nitric oxide moleculs	Guo CX, Ng SR, Khoo SY, Zheng X, Chen P, Li CM	*ACS Nano*, 6:6944–6951, 2012	146
Recent progress in photoswitchable supramolecular self-assembling systems	Yao X, Li T, Wang J, Ma X, Tian H	*Adv. Opt. Mater.*, 4:1322–1349, 2016	133
Controlling supramolecular chirality of two-component hydrogels by J- and H-aggregation of building blocks	Liu G, Sheng J, Wu H, Yang C, Yang G, Li Y, Ganguly R, Zhu L, Zhao Y	*J. Amer. Chem. Soc.*, 140 :6467–6473, 2018	117

nanostructured systems are those that comprise ways to utilize nanoparticles, restrained largely to traditional categories in engineering. Indeed, the recent literature appears to show a shift in the publication of nanostructured systems toward biotechnological application categories.

This shift toward biotechnological applications allows us to better define a smart nanostructured system as one that is designed to solve a specific problem, using interdisciplinary skills to control the utilization of objects that have dimensions on the order of nanometers. Such nanostructured systems could include the utilization of nanoparticles. However, it shows that a smart nanostructured system is better considered as a problem-driven solution, particularly in biotechnology, relying on an engineering approach with a strong basis in science.

Nanostructured Systems for Biomedical and Biological Applications

In the application of a smart nanostructured system to solve a biological problem, we can further consider these systems in the context of engineering that is based strongly in science that may not be exclusively physics. We can illustrate this point with examples of implantable systems that are required to exchange materials or energy with the body. We can define these to be nanostructured symbiotic systems [2].

A symbiotic system comprises an implanted medical device that needs to be designed with an intended two-way duplex communication of energy and materials

Table 2.2 The 12 articles published in 2022 that were identified from a literature search of Wos using the search terms *ALL = (biological engineering) AND (ALL = (smart systems)) AND ALL = (nanostructur*)*. The literature search retrieved 121 items

Article title	Authors	Reference
Enzyme-based mesoporous nanomotors with near-infrared optical brakes	Liu MC, Liang Z, Zhao T, Ma Y, Zhou Q, Ibrahim YS, Elzatahry AA, Li X, Zhao D	*J. Amer. Chem. Soc.*, 144:3892–3901, 2022
A robust combination of dysprosium vanadate/halloysite nanotubes: the elecrochemical system for dimetridazole detection	Kokulnathan T, Chen TW, Chen SM, Ahmed F, Hasan MZ, Bilgrami AL, Kumar S	*Mater. Today Chem.*, 24:100890, 2022
Surface-enhanced Raman spectroscopy substrates for food safety and quality analysis	Nilghaz A, Mousavi SM, Amiri A, Tian J, Cao R, Wang X	*J. Agric. Food Chem.*, 70:5463–5476, 2022
Stimuli-responsive liposomal nanoformulations in cancer therapy: pre-clinical & clinical approaches	Ashrafizadeh M, Delfi M, Zarrabi A, Bigham A, Sharifi A, Rabiee N, Paiva-Santos AC, Kumar AP, Tan SC, Hushmandi K, Ren J Zare EN, Makvandi P	*J. Contr. Rel.*, 351:50–80, 2022
Combining confocal microscopy, dSTORM, and mass spectroscopy to unveil the evolution of the protein corona associated with nanostructured lipid carriers during blood-brain barrier crossing	Battaglini M, Feiner N, Tapeinos C, De Pasquale D, Pucci C, Marino A, Bartolucci M, Petretto A, Albertazzi L, Ciofani G	*Nanoscale*, 14:13292–13307, 2022
Waste-free entropy-driven DNA nanomachine for smartly designed photoelectrochemical biosensing of micoRNA-155	Chen Y, Xu L, Xu Q, Wu Y, Li J, Li H	*Biosens. Bioelectr.*, 215:114569, 2022
Coupled multiphysics modelling of sensors for chemical, biomedical, and environmental applications with focus on smart materials and low-dimension nanostructures	Singh S, Melnik R	*Chemosensors*, 10:157, 2022
Water treatment using stimuli-responsive polymers	Abousalman-Rezvani Z, Roghani-Mamaqani H, Riazi H, Abousalman-Rezvani O	*Polym. Chem.*, 13:5940–5946, 2022
Decoding optical spectra with neural networks to monitor the elimination of carbon nanoagents from the body	Sarmanova O, Laptinskiy K, Burikov S, Dolenko S, Trushina D, Dolenko T	*Opt. Mem. Neural Netw.*, 31:256–265, 2022
Microsized electrochemical energy storage devices and their fabrication techniques for portable applications	Bassyouni Z, Allagui A, Abou Ziki JD	*Adv. Mater. Technol.*, 2200459, 2022
Multifunctional mica extract coated CuO nanoparticles with antimicrobial and dopamine sensing activities: a dual electrochemical-smartphone colorimetric detection system	Ilgar M, Baytemir G, Tasaltin N, Guelluelue S, Yesilyurt IS, Karakus S	*J. Photochem. Photobiol. A*, 43:114075, 2022

(continued)

Table 2.2 (continued)

Article title	Authors	Reference
Bidirectional phase transformation of supramolecular networks using two molecular signals	Cui D, Liu CH, Rosei F, Perepichka DF	*ACS Nano*, 16:1560–1566, 2022

Table 2.3 The WoS disciplines for the top 14 publications with more than 100 citations from the literature described in Table 2.1

Major category (WoS)	Main WoS discipline for highly cited publications				
Chemistry	Chemistry Multidisciplinary	Chemistry Physical	Chemistry Medicinal		
Physics	Physics Applied	Optics	Physics Condensed Matter		
Engineering and Materials	Materials Science Multidisciplinary	Engineering Electrical Electronic	Automation Control Systems	Engineering Mechanical	Robotics
Nanotechnology	Nanoscience Nanotechnology				
Biotechnology	Pharmacology Pharmacy				
Polymers					
Other	Multidisciplinary Sciences				

with the body. We can define a new kind of medical device that is capable of duplex communication and defined as symbiotic device (a "symbio-bot") [2] . Such a concept of a "symbio-bot" extends the existing concepts of what is understood as a biomaterial and biocompatibility, to provide a definition that includes the notion of the intended functional use of the implanted medical device. Such a functional definition includes the important needs for an implanted medical device to be a biocompatible system that is intended and designed to have a two-way (duplex) communication of materials with the body. Such a duplex communicating systems include biofuel cells or open-loop feedback devices where a molecule from the body is utilized by the duplex communicating system to produce a different material (e.g., molecule or energy) for use by the body. Such a duplex level of communication is considered to be a necessary characteristic of symbiotic implanted medical devices that are designed to function for long periods of time inside the body to restore and assist the function of the body.

We now reach the point where we can consider that a smart nanostructured system can truly benefit from combining components of biology with synthetic materials. These components of biology include proteins and lipids that provide biological membrane transport capabilities for the interface of the implanted medical device with the body. Such a system could be termed a "symbio-bot." Now we can

Table 2.4 The WoS disciplines for the 12 articles published in 2022 from the literature described in Table 2.2

Major category (WoS)	Main WoS disciplines for publications in 2022				
Chemistry	Chemistry Multidisciplinary	Chemistry Physical	Chemistry Analytical	Electrochemistry	Chemistry Applied
Physics	Physics Applied	Optics	Biophysics		
Engineering and Materials	Materials Science Multidisciplinary	Instruments Instrumentation			
Nanotechnology	Nanoscience Nanotechnology				
Biotechnology	Pharmacology Pharmacy	Biotechnology Applied Microbiology	Agriculture Multidisciplinary	Food Science Technology	
Polymers	Polymer Science				
Other					

introduce at least four advantages of combining biological components with synthetic materials in smart nanostructured systems to create a "symbio-bot" as being (i) the capability for self-assembly, (ii) the easy scalability that follows from using self-assembly, (iii) easy assembly into 3D structures, and (iv) up to 1000-fold less energy requirements for switching functions compared to state-of-the-art ENODe systems. We will develop these points further in the following sections.

Role of Biological Engineering in the Assembly of Smart Nanostructured Systems

At this point, it is worth to consider a rational approach for the design, assembly, and engineering of "symbio-bots." This rational approach is introduced here as the discipline of "biological engineering." To think about a role for biological engineering, we first need to be clear what we mean by biological engineering. Even if the discussions of this chapter are in the context of assembling nanostructured systems, there are a wide range of applicable descriptions used for the discipline of biological engineering. For example, a search of publications in the Clarivate® Web of Science (WoS) using the terms ALL = (biological engineering) AND (ALL = (bionanotechnology OR nanobiotechnology)) AND ALL = (system*) found 449 published articles. Those articles utilized a large range of words to describe biological engineering in this context, and the 50 most frequently used words are illustrated in Fig. 2.2.

There is a similarity between the WordClouds illustrated in Figs. 2.1 and 2.2, most likely due to the inclusion of the search term *"biological engineering"* for both. Nonetheless, there are subtle but important differences between those two WordClouds. The frequently used words in the WordCloud of Fig. 2.2 provide an insight into the broad range of materials and research outcomes that emanate from research that includes systems that relate to bionanotechnology/nanobiotechnology.

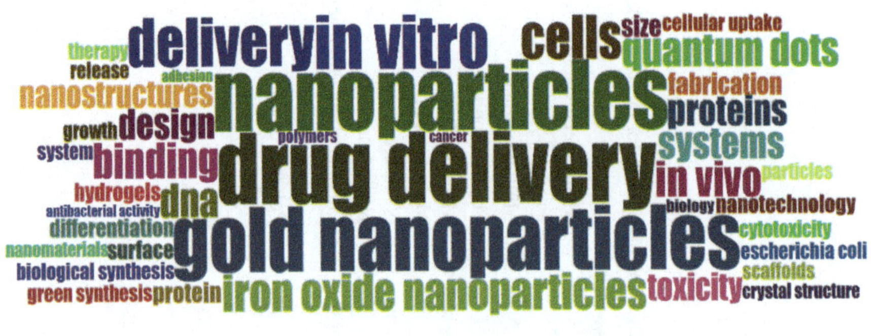

Fig. 2.2 WordCloud of the 50 most frequent words from a bibliometric analysis of the literature in the WoS using the search terms *ALL = (biological engineering) AND (ALL = (bionanotechnology OR nanobiotechnology)) AND ALL = (system*)*. The analysis was performed using the "Bibliometrix" library of RStudio v1.2.5033 [1]

It is interesting to note that the terms *(bionanotechnology OR nanobiotechnology)* imply for most authors in the literature that these smart systems use nanoparticles. This bias is apparent if we compare the WordCloud of Fig. 2.2 to that for a smart system that is nanostructured, as shown in Fig. 2.1. There remains a predominate concept that nanoparticles are the key common feature of a smart system that is nanostructured.

However, neither of the WordClouds in Figs. 2.1 nor 2.2 provide a sense of what is the essence of biological engineering as a discipline. One may ask why is it necessary to define a discipline of biological engineering, when it is apparent that scientists developing nanostructured systems rely on many different disciplines to create research outcomes that utilize nanoparticles. Defining the discipline of biological engineering applied to nanostructured systems is important for the conceptualization and realization for the type of human and material resources that are required to conduct research in the field of nanostructured systems.

As early as the 1930s, biological engineering was used to describe the incorporation of biological knowledge into traditional engineering. In 1970, the term biological engineering was introduced formally and intended to provide an integrated approach to combine engineering with biological systems to move beyond single disciplinary areas such as medicine, agriculture, or fermentation engineering. During the 1980s and 1990s, biological engineering education, philosophy, and practice continued to grow and mature with engineers from agricultural, food, chemical, and biomedical backgrounds increasing to recognize the underlying scientific basis of biology as the driving force for their engineering. Compared to the traditional applications-driven engineering disciplines, perhaps the most prominent aspect that defined this maturing discipline of biological engineering was to let the engineering design process be driven by the unique properties of a biological system (e.g., nonlinearity of system dynamics, time-variant behavior) [3]. Those roots for biological engineering placed it as a science-based discipline in engineering that was not based on the science of physics, such as mechanical engineering or electrical engineering. Moreover, neither was biological engineering an industrial applications-based discipline, such as for mining engineering, power engineering, or agricultural engineering [4]. Having said that, one can further clarify the typology of engineering disciplines based on a different applications-focus that incorporates traditional engineering disciplines with some aspect of a living system [5]. This typology is summarized in Table 2.5.

This diversity of examples contains a common theme, which is the focus on developing externally engineered systems/environments that interact with a living system to alter its behavior/properties. A more fundamental definition of biological engineering would focus on the alteration of existing internal biological processes that control physical attributes/behaviors. Novel advances in micro- and nanotechnologies with electrical/mechanical systems provide the opportunity to work at cellular and molecular scales [5].

For the development of smart nanosctructured systems, we can consider that the discipline of biological engineering will be further differentiated from other engineering disciplines by specializing on the usage of biological elements as

Table 2.5 Typology of engineering disciplines that have some relationship to biological engineering [5]

Engineering discipline	Focus of applications
Civil engineering	Waste treatment remediation
Electrical/computing engineering	Hybrid electronic/living system interfaces and sensors
Chemical engineering	Expertise with fermentations
Agricultural engineering	Food processing
Biomedical engineering	Physical and energetic interfaces between physical engineered systems and human living systems

components in the design of the smart nanostructured systems. Such biological components will include lipids, proteins, enzymes, and other molecules that mimic their natural function in a biological system. Such biologically engineered smart nanostructured systems may include nanoparticles, but it is not the nanoparticles per se that defines the assembly of a smart nanostructured system utilizing biological engineering.

Naturally Smart Nanostructured Systems in Biology

This section will focus on biological systems that are capable of providing bioinspiration for the assembly of smart nanostructured systems. For this, our approach here is to consider the cell as an individual component of a larger engineering system. In that context, an individual cell is the smallest functional unit that is capable of working as a stand-alone system, but each such functional unit contributes to the overall function of a complete biological system. Essentially, the cells can function individually, but the communication together provides the biological function of a complete tissue in the body. The membrane of each cell provides the interface for each of these functional units within the overall system of cells in the tissue. The membrane proteins have important roles to provide the output effectors for each cell to communicate with its neighboring cells. Concomitantly, the membrane proteins also provide the neighboring cells with inputs that assist the communication between cells to support the coordinated overall function of the tissue.

Clearly, inside each cell, there are complex interactions of genes and molecules with the cell nucleus and other organelles. These complex interactions are essential for the biological life of the cell. Also, essential for biological life is the immune system, comprising the reticulo-endothelial system. The immune system protects the tissues of the body from infection and is thus an essential aspect for maintaining the ongoing health of biological tissues.

It is not possible to engineer life, but it is possible to engineer an individual functioning nanostructured "artificial cell" to mimic the membrane transport functions of a living biological cell [6]. That is, the artificial cell is engineered with the capability of a biological cell for membrane transport communication. If the defining feature of a biological system is the capability of the individual cells to communicate together, then a smart nanostructured system can thus be engineered to mimic the overall function of a biological system. In this context, the biological system responses are the important functional outputs to be mimicked by a smart nanostructured system.

In describing naturally occurring biological systems in this chapter, we will utilize the terminology of a "network" in a similar sense to that of a "system." In a strict engineering usage of these terms, a "network" usually refers to a system that connects two or more computing devices for transmitting and sharing information. The computing devices are connected using either physical wires, including fiber optics, or wireless transmission of signals. A "system" is usually defined as a group of interacting or interrelated elements that act according to a set of rules to form a unified whole to produce a certain outcome.

The interaction between biological cells does not occur via wires, but by various molecules that are secreted by cells and then sensed by receptors on the cell membrane. In excitable cells, such as nerves or muscles, there is also a capacitive coupling between cells that rely on nan-Faradaic interactions of electrical potential. Hence, the communication in a network of cells is due to combined electrochemical forms of transmission. Biological cells are capable of self-assembling into complex 3D structures. This implies that a network of biological cells is also a system, but one in which there may not be an overarching set of predefined rules for the output of the unified whole of cells to produce a certain outcome.

We will explore how different systems of biological cells inspire the inspiration for the biological engineering of smart nanostructured systems.

Networks of Second Messengers Inside Cells That Are Important for Tissue Functions

Each cell of the body comprises a complex network of second messenger molecules that are essential for specific functions of specific cells. Nonetheless, the general function of an intracellular signaling network is illustrated in Fig. 2.3. The second messenger network inside the cell is triggered by the binding of a first-message signal molecule from the extracellular fluid to a protein receptor incorporated in the cell membrane. The first-message molecule can be either naturally occurring (e.g., neurotransmitters, cytokines) or therapeutic molecules to treat disease. This ligand/receptor binding then triggers a cascade of intracellular biochemical events that contribute to an intracellular networked system. The advantage of such a second-messenger signaling network is that there is a biochemical amplification of the initial ligand/receptor first-message so as to ensure that there is an appropriate response from the cell, which can be sustained.

We will utilize the example of the kidney to provide a detailed description of a functioning second messenger network. The kidney is an extremely important tissue

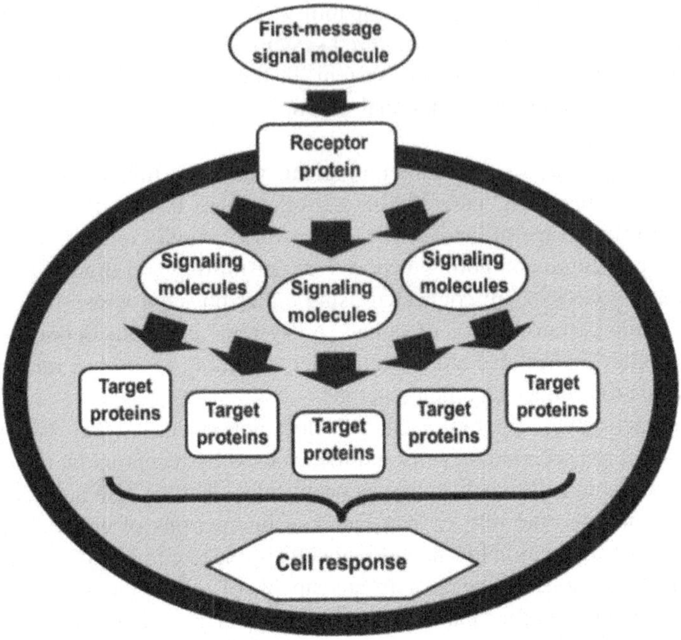

Fig. 2.3 A general system of an intracellular second-messenger signaling network. The thick black line represents the cell membrane that encloses the interior of the cell (colored gray). The receptor protein incorporated in the cell membrane binds a ligand molecule from the extracellular fluid. This binding triggers the release of signaling molecules that are usually the α or βγ subunits of G-proteins associated with the receptor protein. The target proteins inside the cell can include protein kinases, enzymes, or calcium-binding proteins. This type of intracellular signaling network can lead to a multitude of cell responses, which include the activation of ion-transporting proteins, the phosphorylation of proteins, cell contraction, secretion of other molecules, changes in cell metabolism, immune responses, or regulation of gene expression

that utilizes membrane transport to control fluid balance and nutrients/waste in the body. The response/effectors are membrane transport proteins. The kidney serves an excellent model for a smart nanostructured biological system providing the inspiration for the biologically engineered smart nanostructured system.

We describe this kidney system in the context of the cell physiology of clear cell carcinomas (CCC). Such CCC are identified by a clear cytoplasm in neoplastic cells during histological analyses, which is due to histological processing that dissolves the high lipid content of the CCC [7]. The clear-cell morphology of CCC is caused by highly active lipid and glycogen synthesis and deposition in the cytoplasm as an increase in lipid droplets, which reflects a unique metabolic state of the CCC. The CCC constitute the most frequent and metastatic form of kidney malignancy and the most therapy-resistant form of ovarian carcinomas [8]. With regard to cell physiology of clear cell renal cell carcinoma (ccRCC), membrane ion channels provide a cell response following the activation of second-messenger pathways that are stimulated by therapeutic medications used for ccRCC.

Despite being a smaller subset of the total number of proteins encoded by the human genome, membrane-bound proteins provide a major target for pharmaceutical drugs. Uhlén et al. provide an analysis of the total human genome that indicates there are 20,344 putative protein-encoding genes [9]. Notwithstanding that there are differences in protein expression between different tissues and that around two-thirds of protein-encoding genes have splice variants for different cellular locations, there are about 5500 genes that are predicted to encode membrane-associated proteins, with the proportion of specifically identifiable membrane-bound proteins being 14% (2804 genes). Those membrane-bound proteins include ion channels or molecular transporters, enzymes, and anchors for other proteins [9]. Membrane-bound proteins are essential for several important cell functions that include the transmembrane transport of ions and small molecules, signal transduction, and enzymatic activity [10]. Indeed, such membrane-bound proteins provide the targets for more than 60% of the drugs in clinical use [11].

As a clinical example for the role of membrane ion channels in ccRCC, the ATP-gated membrane ion channel (P2X7) is involved in inflammation, apoptosis, and cell proliferation and thus plays a crucial role during oncogenic transformation in various malignancies. In a retrospective clinical study of 273 patients with ccRCC undergoing nephrectomy, it was found that the high intratumoral expression of P2X7 was associated with shortened cancer-specific survival. From that retrospective clinical study, the increased intratumoral expression of P2X7 appeared to be a potential independent adverse prognostic indicator for postoperative cancer-specific survival in ccRCC [12].

P2X receptors are a family of ionotropic ATP-gated ion channels that are equally permeable to Na^+ and K^+ and with significant Ca^{2+} permeability. Of this family, P2X7 channels are unusual in that there are extra amino acids in the intracellular C terminus. P2X7 are key mediators of inflammation and cancer invasion/metastasis. Antagonists to P2X7 are potential anti-metastatic agents. Moreover, the use of P2X7 antagonists has a potential role in other types of kidney failure, since there is an increased expression of P2X7 in renal hypertension, polycystic kidney disease (PKD), and glomerulonephritis [13].

Autosomal dominant polycystic kidney disease (ADPKD) is characterized by multiple renal cysts, hypertension, and a decline in renal function, leading to renal failure in 50% of all AHPDK patients. Cysts can develop in each nephron segment through a combination of aberrant epithelial cell proliferation and abnormal fluid secretion. Cystic fluid is produced by a cAMP-dependent mechanism similar to that found in secretory epithelia. An important step in cAMP-dependent fluid secretion is the activation of the cystic fibrosis transmembrane conductance regulator (CFTR), which secretes chloride into the cyst lumen followed by sodium and water. CFTR is a cAMP-dependent chloride channel that is best known for its role in cystic fibrosis (CF). A drug, VX-809, used to correct CF, was found to reduce the growth of cysts in ADPKD and to increase the expression of CFTR [14].

The role of cation channels in renal cell carcinoma (RCC) was demonstrated by using the natural drug product Englerin A to inhibit the growth of the renal tumor cell line A-498 by activating transient receptor potential cation channels (TRPC4 and TRPC5) [15].

A large proportion of cancer patients fail to respond to conventional chemotherapy because of the intrinsic resistance of their cancer to pro-apoptotic stimuli and/or the acquisition of a multidrug resistant (MDR) phenotype during chronic chemotherapy. A significant proportion of kidney cancers overexpress the alpha-1 subunit of the Na^+-K^+-ATPase as compared to corresponding normal tissues, while colon cancers overexpress the alpha-3 subunit. For example, cardiotonic steroids including cardenolides and bufadienolides have anti-cancer effects but with limited cardiotoxicity. An approach to improving the clinical outcomes of chemotherapy could be to target the specific ion channels and subunits of the Na^+-K^+-ATPase that are overexpressed in cancer cells compared to normal cells [16].

The function of the kidneys depends on the maintenance of an epithelial barrier that is capable of regulating paracellular permeability to solutes and water. The tight junctions (TJ) provide that epithelial barrier function, which can be measured using techniques such as transepithelial resistance (TER) or electrical impedance spectroscopy (EIS) [17]. The control of permeability is attributed to the insertion of discrete paracellular tight junction channels (PTJC) using evidence from elegant biophysical experiments, although at the time of that work it was not known how the TJ proteins could assemble to form PTJC. Such TJ proteins were known to include occludin (3 isoforms), claudin (> 21 orthologs), and junction-associated molecule (JAM), although the stoichiometry of these proteins in a single TJ fibril and their oligomerization in heteromeric and heterotypic assemblies were not known [18].

Using quantitative reverse transcriptase–polymerase chain reaction (rtPCR) and immunohistochemistry, it was shown that claudin-2 mRNA was expressed in the kidney, liver, pancreas, stomach, and small intestine. The highest level of which was detected in the kidney. Also, colorectal cancers (CRCs) express claudin-2 mRNA at high levels [19]. The presence of claudin-2 causes the formation of cation-selective channels that then increase the permeability of the TJ, whereas claudin-1 and claudin-4 decrease the permeability of the TJ [20]. The overexpression of occludin increases cancer cell sensitivity to apoptogenic factors and consequently, lowers tumorigenicity and invasive capacity [21].

The control of TJ permeability is complex, as shown by the experiments that used epidermal growth factor (EGF) extracted from human urine to decrease TJ permeability. This decreased permeability was due to decreased expression of claudin-1 and claudin-2 and increased expression of claudin-4, with the resulting permeability changes being transient despite the continuous presence of EGF. These complex kinetic responses may be understood by considering that EGF also stimulates the secretion of prostaglandins in kidney cells, and it is known that the prostaglandin E_2 inactivates the extracellular-regulated kinases 1/2 (ERK1/2). Thus, prostagladin E_2 could provide a negative feedback to turn off the signal initiated by EGF, by acting through the cross-talk of the intracellular Gαs and ERK1/2 pathways [22, 23].

Subsequently, the scaffolding protein, IQGAP1, was shown to regulate claudin-2 and claudin-4. Claudin-2 forms a cation-selective pore at the TJ and increases the permeability (measured by decreased TER). Claudin-4 reduces the Na^+ permeability of the TJ (measured by increased TER). Moreover, measurements of claudin 4 expression have a prognostic value in several tumor types and in the association between the loss of TJ and metastatic behavior. Taking all that evidence together suggested that IQGAP1 could be a potential therapeutic target for the treatment of advanced cancers [24].

Biological Engineering of Smart Nanostructured Systems

Here we will describe biological engineering in two contexts. The first context is to utilize a living cell and to transform its function using the tools of biology to make genetic manipulations of the protein expression in the cell. The second context is the construction of an artificial cell into which biological proteins are inserted to provide ion transport functions for the artificial cell.

First Context: Genetic Manipulations to Transform the Function of Cells for Therapeutic Benefits

As part of our research into the liver-directed gene therapy of Type I diabetes, we have genetically engineered a human hepatoma cell line (HEPG2ins/g cells) to store and secrete insulin to a glucose stimulus. The engineered cell line (HEPG2ins/g) responds to glucose via intracellular signaling pathways that depend on K_{ATP} ion channels, similar to a normal pancreatic β cell [25]. We used this genetic engineering approach to induce a capability for intracellular pathways to control the function of the induced K_{ATP} ion channels. There are functional K_{ATP} channels expressed in the transfected liver cell-line HEPG2ins/g. These channels are activated only in the parent HEPG2 cell line by pharmacological stimulation. We have demonstrated, by using patch-clamp electrophysiology, confocal laser scanning microscopy with labeling of insulin-containing secretory granules, measurement of intracellular Ca^{2+} levels, and biochemical studies of glucose-stimulated insulin secretion, that the HEPGins/g cells secrete insulin in response to glucose in a manner consistent with the classical triggering pathway present in pancreatic β cells. There is potential value for the HEPG2ins/g cells to be considered as a genetically engineered "artificial β cell" in the treatment of Type I diabetes.

Biological Control of Insulin Section in Pancreatic β-Cells

Type I diabetes or IDDM is caused by the autoimmune destruction of pancreatic β cells [26]. Current treatment of the disease requires daily injections of insulin to control blood glucose levels. Results from The Diabetes Control and Complications Trial Research Group [27] show that the onset of diabetic complications, which greatly reduces the quality and longevity of life in IDDM patients, is reduced by tight glucose control. In a normal pancreatic β cell, a rise in extracellular glucose initiates the inhibition of ATP-sensitive potassium channels (K_{ATP}), which leads to depolarization, influx of extracellular Ca^{2+} ions, induction of a rise in $[Ca^{2+}]i$ from intracellular stores, exocytosis, and secretion of insulin [28–31]. The K_{ATP} channel has been cloned and found to be a complex of a K^+ channel (Kir 6.2) and an ATP binding cassette protein (SUR1) that functions as a high-affinity receptor for sulphonylureas [32–35].

Although potassium channels have been described in many cell types, apart from pancreatic islets, including skeletal muscle, cardiac, and vascular myocytes; neurons; and renal epithelial cells [36–40]; characterization of the role of K^+ channels in hepatocytes has been limited. Henderson et al. [41] described the presence of inwardly rectifying K^+ channels in primary rat hepatocytes that were not affected by voltage or Ca^{2+} stimulation. Capiod and Ogden [42] described the presence of Ca^{2+}-activated and delayed-rectifier K^+ currents, and Lidofsky [43] reported the presence of Ca^{2+}-sensitive and cAMP-dependent K^+ channels in HTC rat hepatoma cells.

Genetic Engineering of "an Artificial β Cell"

Glucose control could theoretically be improved by genetically engineering "an artificial β cell" that is capable of synthesizing, storing, and secreting insulin in response to metabolic signals. In pursuit of this goal, hepatocytes have been shown to be a suitable target cell [44–49]. As the starting cell for our genetic engineering, we chose hepatocytes (liver cells) since these cells are known to play a crucial role in intermediary metabolism and synthesis of proteins in the liver. Most importantly liver cells express the high-capacity glucose transporter GLUT2 [50] and the glucose phosphorylation enzyme glucokinase [51], which comprise the key elements of the "glucose sensing system," which regulates insulin secretion from pancreatic β cells in response to small external nutrient changes. The insertion of insulin cDNA into a human hepatoma cell line (HEPG2) that lacked native GLUT 2 expression to produce the cell-line HEPG2ins resulted in both the synthesis, storage, and release of insulin to β cell secretagogues, but not to glucose [44]. A second insertion of the glucose transporter GLUT 2 resulted in near physiological release of insulin to glucose and other stimuli in the doubly transfected HEPG2ins/g cells [45].

As a smart nanostructured system, the novelty of the HEPG2ins/g cells is that dual expression of the glucose transporter GLUT 2 and the insulin gene has induced functional K_{ATP} channels in the resting cells. Transfection of the parent HEPG2 cell-line with both the GLUT2 transporter and insulin cDNA resulted in cells (HEPG2ins/g) that stored insulin in secretory granules and secreted insulin in response to glucose stimulation, at near physiological levels [45]. That genetic manipulation created a smart nanostructured system that expressed functional K_{ATP} channels in the resting HEPG2ins/g cells to provide a mechanism to trigger insulin secretion in response to stimulation of the cells by glucose [52, 53].

At this point, it is important to highlight the steps in the genetic manipulations to show how this "artificial β cell" was developed. Patch-clamp electrophysiology was used to measure that the doubly transfected HEPG2ins/g cells, in which cDNA for insulin and GLUT2 genes had been inserted, had a more negative membrane potential (−68.9 ± 7.1 mV) than either the untransfected HEPG2 cells (−18.2 ± 5.8 mV) or the HEPG2ins cells that had only the insulin gene inserted (−45.3 ± 4.4 mV). Furthermore, the increased membrane potential was due to the functional activity of K_{ATP} ion channels induced in the HEPG2ins/g cells (Fig. 2.4).

The activity of the K_{ATP} channels was inhibited by cAMP, ATP (Fig. 2.5), and glucose (Fig. 2.6).

To provide a pharmacological means to control the K_{ATP} channels during experiments, we also demonstrated that the K_{ATP} channels were activated by diazoxide and inhibited by glibenclamide (Fig. 2.7).

The activation by diazoxide would be expected to inhibit the glucose-stimulated cell depolarization and thus provide an important control for the experiments.

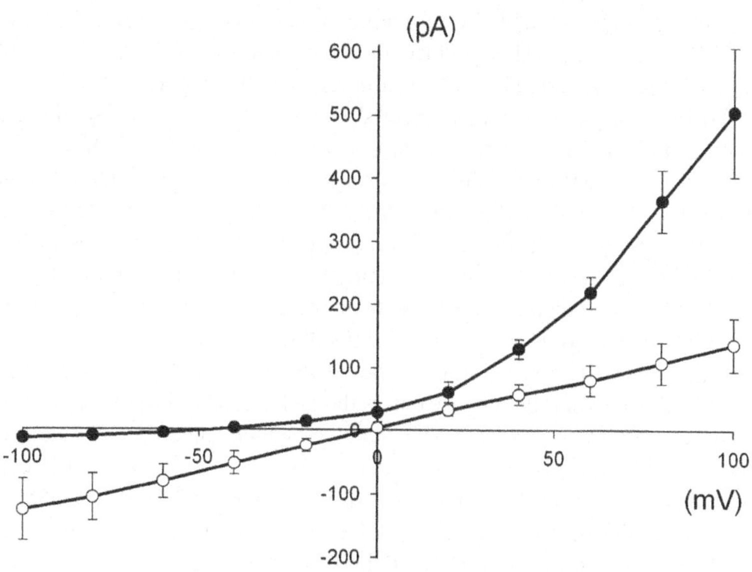

Fig. 2.4 Summary of whole-cell I-V curves for the samples of HEPG2ins/g ($n = 3$) and HEPG2 cells ($n = 5$). Vertical bars indicate SEM. The position of the reversal potential ($E_R \approx -50$ mV) for the HEPG2ins/g cells indicates a potassium selective current, whereas the current for the HEPG2 cells is non-selective ($E_R \approx 0$ mV). (Figure reproduced from [53])

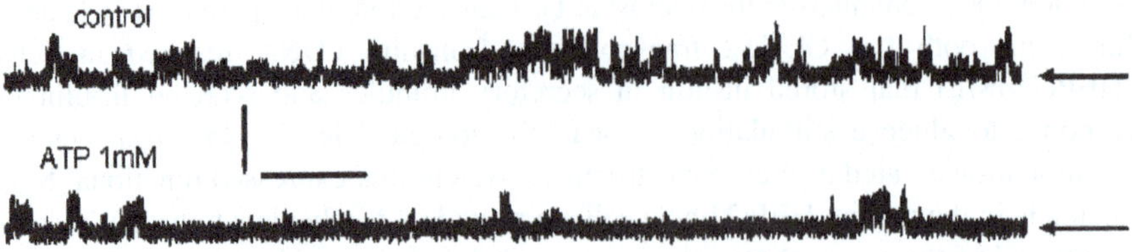

Fig. 2.5 Representative single-channel recordings of K_{ATP} activity in an inside-out membrane patch of HEPG2ins/g cell (*n* = 8 experiments in total). Closed channel level is shown by the horizontal arrows, and open channels are upward deflections. ATP (1 mM) reduced the frequency of open channel activity. Transmembrane potential = +40 mV. The vertical scale-bar represents 10 pA, and the horizontal scale-bar represents 200 ms. (Figure reproduced from [53])

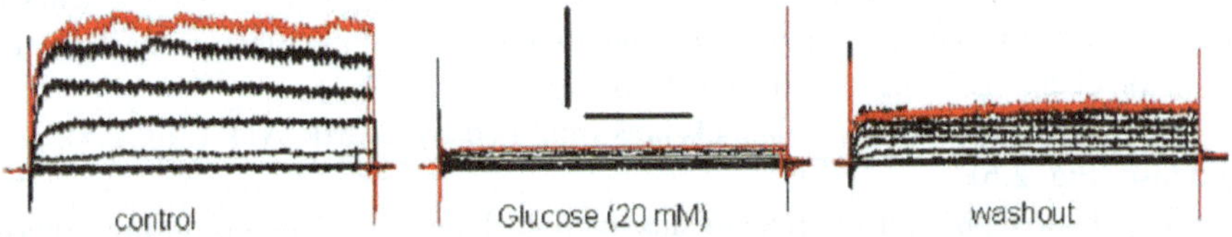

Fig. 2.6 Sensitivity of K_{ATP} ion current to high-glucose concentration. Representative whole cell patch clamp currents of the HEPG2ins/g cells (*n* = 4 experiments in total). Glucose (20 mM) was added to the superfusing solution and reversibly inhibited the K_{ATP} currents. The vertical scale-bar represents 200 pA, and the horizontal scale-bar represents 200 ms. (Figure reproduced from [53])

Furthermore, we confirmed by Western blotting that the K_{ATP} channels were induced in the HEPG2ins/g cells. This confirmation was achieved with the use of an antibody against the conducting pathway of the K_{ATP} channels, Kir6.2 (Fig. 2.8).

The HEPG2ins/g cell ("artificial β cells") also responded to stimulation by glucose with an increase in the intracellular Ca^{2+} concentration (Fig. 2.9).

Ca^{2+} is an important intracellular messenger that is needed to trigger the secretion of insulin. We confirmed that the HEPG2ins/g cells were capable of secreting inclusion in response to glucose, which was controlled by the activity of the induced K_{ATP} channels (Fig. 2.10). Note that activating K_{ATP} channels leads to hyperpolarization of the cell, which then does not lead to an increase in intracellular Ca^{2+} concentration, and hence, there is an inhibition of insulin secretion.

Thus, this nanostructured system of the HEPG2ins/g cells follows the classical triggering pathway whereby glucose enters the cell and there is a rise in the ATP-to-ADP ratio leading to closure of KATP channels, which causes membrane depolarization and subsequent activation of insulin secretion by exocytosis of secretory granules [54]. The combined transfection of the GLUT2 and insulin genes induces the activation of functional K_{ATP} channels that do not require pharmacological stimulation to generate currents in this "artificial β cell."

The function of this nanostructured "artificial β cell" is illustrated in Fig. 2.11. The expression of functional K_{ATP} channels in the resting HEP G2ins/g cells provides a mechanism for the rapid triggering of insulin secretion. Glucose in the

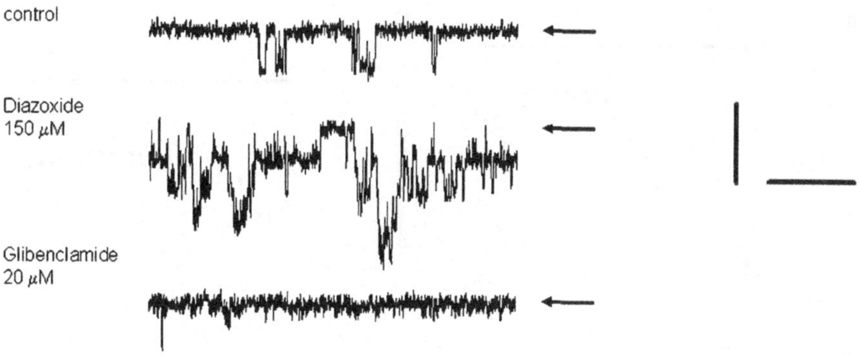

Fig. 2.7 Single-channel activity of K_{ATP} in an outside-out patch was increased by the channel opener diazoxide (150 μM) and inhibited by the channel blocker glibenclamide (20 μM). Closed channel level is indicated by the horizontal arrows, and open levels are downward deflections. Transmembrane potential = -40 mV. The vertical scale-bar represents 5 pA, and the horizontal scale-bar represents 100 ms. (Figure reproduced from [53])

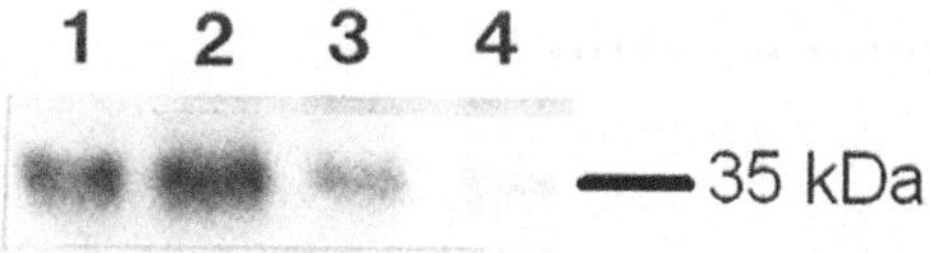

Fig. 2.8 Qualitative Western blot analysis for Kir6.2 in HEPG2ins/g (lane 1), MIN 6 (lane 2), HEPG2ins (lane 3), and HEPG2 cells (lane 4). The molecular weight of Kir6.2 is 35 kDa. The MIN 6 cells (positive control) had the strongest staining for Kir6.2. There was a strong presence of Kir6.2 in the HEPG2ins/g cells. There was weaker staining for Kir6.2 in the HEPG2ins cells and almost no staining for Kir6.2 in the HEPG2 cells. This pattern of Kir6.2 expression follows the electrophysiology data, which indicated that the functional presence of the KATP channels was greatest in the HEPG2ins/g cells and absent in the HEPG2 cells. (Figure reproduced from [53])

extracellular environment is sensed and transported across the membrane by the facilitative glucose transporter GLUT2 and converted to glucose-6-phosphate by glucokinase; there is a rise in the ATP:ADP ratio leading to the closure of K_{ATP} channels that leads to cell depolarization, influx of extracellular Ca^{2+} to increase the intracellualr concentration, increased release of Ca^{2+} from intracellular stores, exocytosis, and secretion of insulin. As for the situation of a normal pancreatic β-cell it seems likely this increase in Ca^{2+} would also have a wide range of effects including the activation of glycerol phosphate dehydrogenase and other mitochondrial enzymes and activation of protein kinases and phospholipase C. The subunits that comprise the K_{ATP} channel in natural pancreatic β-cells are SUR1 and KIR 6.2. The subunits comprising the K_{ATP} channel in the parent HEP G2 cells were reported to be SUR2 and Kir 6.2, whose mRNA expression was unaltered by glibenclamide. In this "artificial β cell," following overexpression of the insulin and GLUT 2 cDNA, the HEPG2ins/g cells express K_{ATP} channels that are blocked by both tolbutamide and glibenclamide and activated by diazoxide. This pharmacology suggests that the

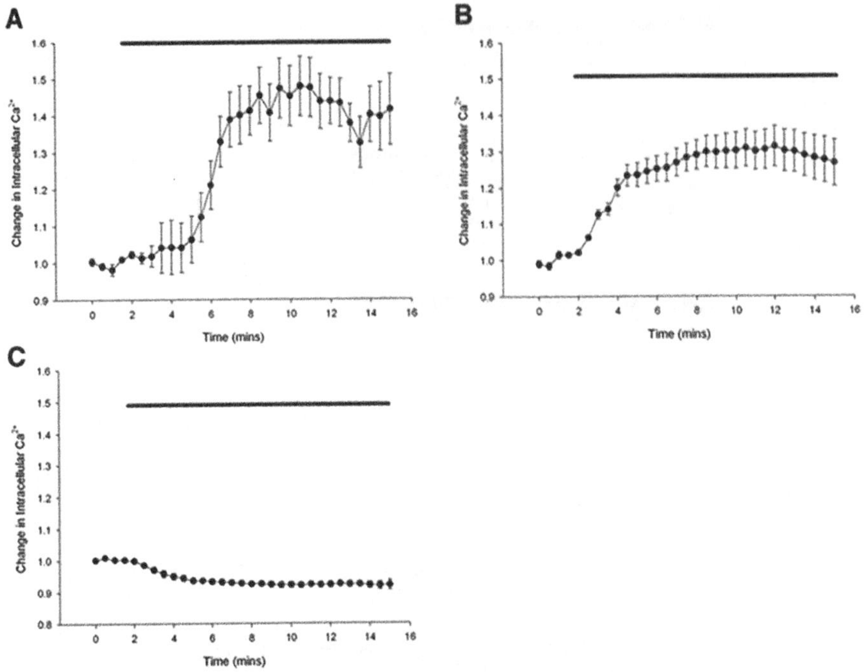

Fig. 2.9 Intracellular free Ca2+ changes in HEPG2 and HEPG2ins/g cells induced by glucose and the K_{ATP} channel antagonist glibenclamide. The vertical bars on each data point indicate SE where this is larger than the symbol. (**a**) Glucose 20 mM stimulated an increase in the intracellular free Ca2+ level in HEPG2ins/g cells ($n = 12$). (**b**) Glibenclamide 100 μM also stimulated an increase in the intracellular free Ca2+ level in HEPG2ins/g cells ($n = 9$). (**c**) Glucose 20 mM failed to stimulate an increase in intracellular Ca2+ level in the HEPG2 cells ($n = 10$). The horizontal bars indicate the time of application of glucose or glibenclamide. (Figure reproduced from [53])

expression of the insulin and GLUT 2 genes in these cells possibly modified the SUR2 subunit of the existing channels to enable the K_{ATP} channels to be functional. This "artificial β cell" supports the possibility to engineer liver cells to mimic the physiology of pancreatic β cells and the eventual use for the treatment of type I diabetes.

Second Context: Ion Transport Proteins Are the Smart Components for Nanostructured Systems

We have seen how a living cell can be used as the basis for genetically engineered nanostructured system to achieve a specific and different purpose compared to the original function of the living cell. However, for a biological engineering approach, we need to consider how to construct a nanostructured system that does not need to

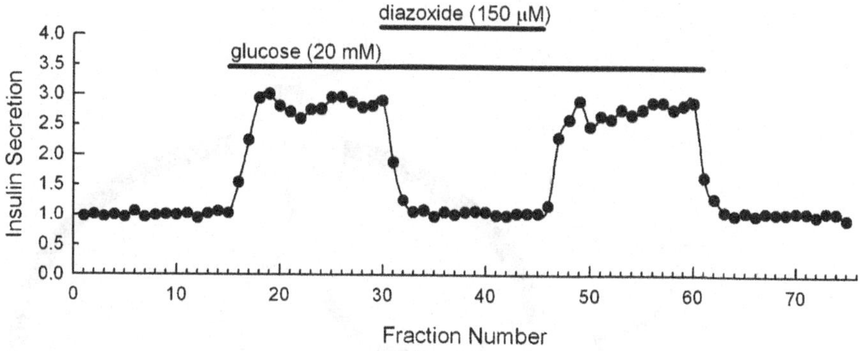

Fig. 2.10 Static stimulation of insulin secretion in cultures of HEPG2ins/g cells. Data have been normalized to the average insulin secretion measured in samples between 0 and 15 min (1.00 ± 0.001 pmoles insulin/108 cells), during which cells were perfused with basal medium without any stimuli. The y-axis shows the increase in insulin secretion as a multiple of the normalized basal secretion (= 1.0). Data are representatives of three independent experiments. (Figure reproduced from [53])

modify an existing living cell. In other words, we need to utilize the principles we have learned from the genetic engineering approach but apply these to synthetic nanostructured systems using a biological engineering approach. We will illustrate such a biological engineering approach using the example of an artificial cell that was constructed using synthetic and biological components [6].

The synthetic component for that artificial cell was a hollow PSS/PAH polyelectrolyte polymer capsule of 6 μm in diameter (Fig. 2.12). The nanostructured biological components comprised a lipid bilayer membrane to enclose the hollow polyelectrolyte capsule, and ion channels incorporated into the lipid bilayer membrane. This complete system provides a structurally stable microcapsular system and addresses the concern expressed previously about liposomes that incorporated a porin to facilitate membrane permeability [55].

The ~20 nm thick wall of the hollow polyelectrolyte capsule allows the transfer of transition metal complexes with overall large positive or negative charges as well as neutral species while providing enough structural strength to maintain a spherical shape when coated with a lipid bilayer. The lipid coating provides a seal, containing both large and small solute species within the microcapsule. The incorporation of the engineered ion channel, gramicidin (bis-gA), provides a platform that can be used to effectively control transport of molecules from the microcapsule. This engineered micro/nanosystem has potential applications in physiologically relevant delivery or sensing applications. Indeed, we have reported an algorithm and method that would enable such a system to be used efficiently in a biosensing application [56]. The opening of bis-gA suggests that the bis-gA ion channels incorporated rapidly into the lipid bilayer around the microcapsules, providing functional and gateable pathways to control ion transport across the microcapsules.

In this type of "artificial cell" as a smart nanostructured system, the intriguing aspects of variations in the kinetics of the gating of incorporated

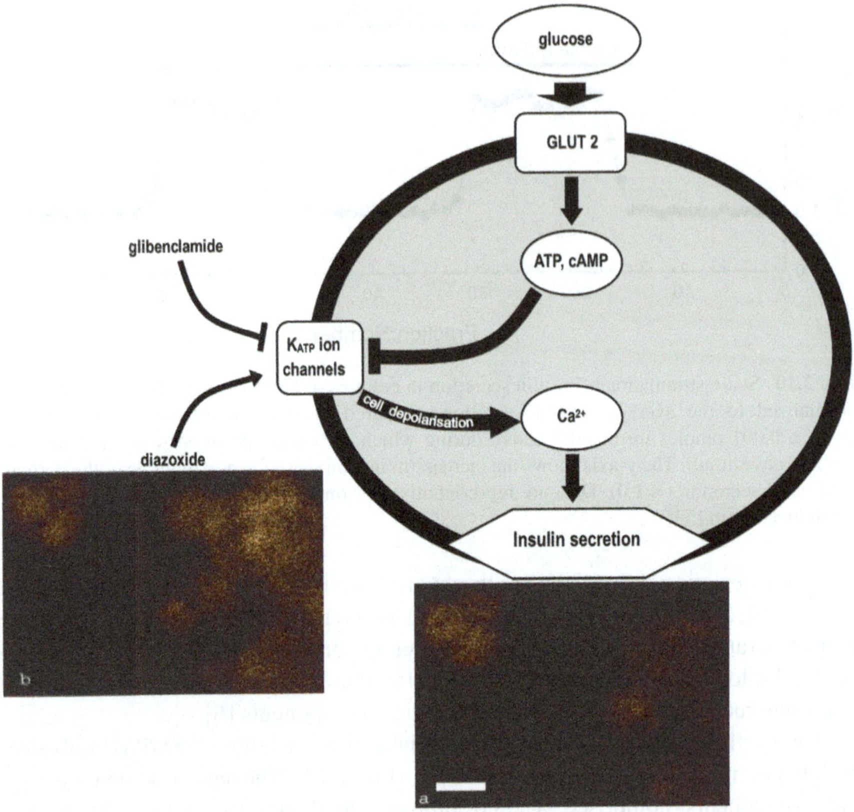

Fig. 2.11 Schematic diagram of the biochemical pathway of glucose-stimulated insulin secretion from HEP G2ins/g cells. A rise in extracellular glucose initiates the inhibition of K_{ATP} channels, which leads to cell depolarization, exocytosis, and secretion of insulin. In normal pancreatic β cells, depolarization is due to influx of Ca^{2+} ions. The secretion of insulin is illustrated by the confocal microscopy images (**a**) and (**b**) of Zinquin-E fluorescent labeling of secretory granules in HEPG2ins/g cells [53]. Image (**a**) shows Zinquin staining of secretory granules in HEP Gins/g cells superfused with 5 mM glucose in DMEM medium (control). Image (**b**) shows the same HEPGins/g cells as in (**a**), after changing the superfusing solution to one containing 150 μM diazoxide in DMEM medium followed by the addition of 20 mM glucose in DMEM medium 15 min later. The image was taken after 45 mins' exposure to 20 mM glucose medium. The greater punctate staining in (**a**) indicates that the K_{ATP} channel activator (diazoxide) has inhibited the glucose-stimulated secretion of insulin-containing granules in the HEP Gins/g cells despite prolonged exposure to 20 mM glucose

membrane-spanning bis-gA ion channels provide the possibility to use this "artificial cell" to gain further fundamental insights into the mechanisms underlying instability in the lipid coating of polyelectrolyte supports. Measuring the kinetics of channel open/close activity could be a valuable addition to techniques such as colloidal force spectroscopy [57] in enhancing an understanding of the mechanisms of interaction between lipid bilayers and polyelectrolyte supports. Notwithstanding,

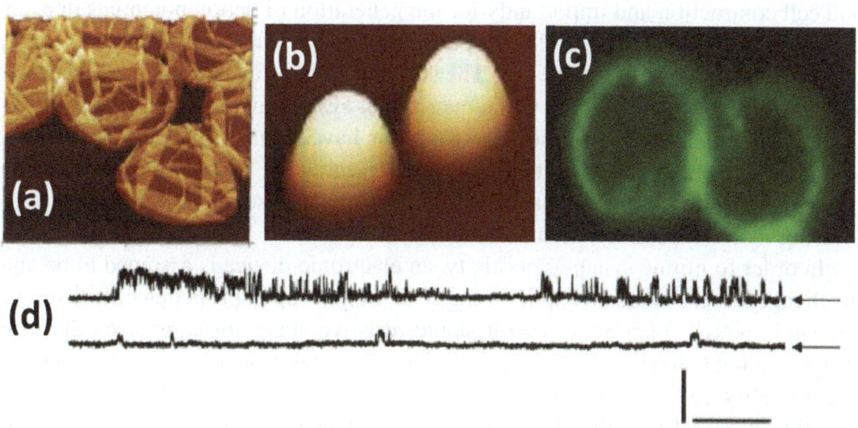

Fig. 2.12 Structures to form an artificial cell. (**a**) 3D rendered AFM images of hollow PAH/PSS polyelectrolyte microcapsules showing a folded and layered structure in an aqueous environment. (**b**) The microcapsules with a lipid coating do not collapse and maintain a spherical structure in an aqueous environment. (**c**) Confocal microscopy images of hollow lipid-coated microcapsules loaded with dextran-FITC by adding 10 μL of 10 kDa dextran-FITC to 1 mL of a 1 mg/mL. After washing, the lipid-coated microcapsules retained the fluorescence, which showed the lipid membrane formed an effective insulating barrier to seal the microcapsules. (**d**) Continuous 10 second patch-clamp recordings of the gating of the bis-gA channels incorporated in the lipid coating of the capsules by MBC (10 μM). MBC is an acronym for the compound *N,N0-dimethyl-N-(2-[2-(methyl-4-[1,1,3,3-tetramethylbutyl]phenoxy)ethoxy]ethyl)benzylammonium chloride*. The upper trace is the control condition, where the upward deflections indicate the ion current permeating across the wall of the capsules to the opening of the bis-gA channels incorporated in the lipid coating. The lower trace is recorded 2 min after addition of the drug to the solution bathing the capsules. The MBC blocked the opening of bis-gA and hence the ion permeation across the capsule wall, as shown by the reduction in the upward deflections. The horizontal arrow indicates the nonconducting condition of the bis-gA channels where there is zero current recorded. The horizontal scale-bar represents 1 s and the vertical scale-bar represents 5 pA. The four panels were compiled from figues in [6]

the "artificial cell system provides a platform for the development of a sealed micro-container for drug delivery or smart sensing that can navigate in the physiological environment and be controlled with physiologically relevant stimuli. Our evidence and formulation of a Hidden Markov Chain Model for automatic detection of changes in the ion channel gating reinforces the potential physiological applications of this "artificial cell" system that incorporates ion channels [56]. The controlled transport in this "artificial cell" system could be tuned for selective release biomimetically by controlling the gating of incorporated ion channels. This system provides a platform for the creation of "smart" biomimetic delivery vessels for the effective and selective therapeutic delivery and targeting of drugs.

Membrane transport proteins also provide the ion gradients that is the basis for cells to develop differences in electrical potentials across their membranes. Such gradients provide the energy for many functions, including a triggering mechanism for the inflow of Ca^{2+} ions through voltage-gated ion channels into muscle cells to aid cell contraction and importantly for the generation of action potentials in excitable cell membranes. A good example for a body tissue that contains excitable membranes is the retina of the eye. The state-of-the-art in electronic neuromorphic visual systems aims to mimic the human visual system and to reproduce the complete functionality of a synapse. Those devices have some limitations in achieving the typical synaptic process that underlies learning in biological systems (e.g., activity-dependent synaptic plasticity), which may be resolved by biological engineering with ion channels in an "artificial cell" system.

In order to mimic synaptic plasticity, an electronic device is assumed to be able to change its resistance (synaptic strength, or weight) upon proper electrical stimuli (synaptic activity) showing several stable resistive states throughout its dynamic range (analog behavior). The electronic component called a memristor provides that functionality [58, 59]. Indeed, memristors provide the CMOS-based electronic building blocks for artificial devices capable of pattern-learning and recognition [60–63]. In 2018, the use of multiple memristive devices made a significant advance in creating the basis for building the next-generation of intelligent computing systems [63]. Despite the considerable advances in electronic design to develop the memristor, those electronic devices still consume significantly greater energy to perform neuromorphic computing functions and remain much larger in size than biological neurons and synapses.

Most electronic memristors are volatile in the sense that, after the power is turned off, the device reverts to a fixed memristance (or memductance) state. However, an ideal memristor is nonvolatile and remembers its most recent memristance (or memductance) state when the power is turned off. This is due to the properties of internal state-variables in the memristor, which are physical parameters such as temperature, pressure, impurity concentration, chemical moiety, etc., that do not depend on any external variables such as the voltage or current associated with another device. The internal state-variables render the memristor with a strictly monotonically increasing transconductance constitutive relation, and hence, a nonvolatile memory because the memristor remains in the last memristance (or memductance) value from the continuously tunable slope along its transconductance curve is due to the ability of the memristor to store its transconductance curve. It follows that the ideal memristor is capable of the simplest form of learning, namely, habituation [58]. The internal molecular configuration of membrane ion-conducting proteins defines their ion-conducting properties. Thus, their transconductance is defined by an internal state-dependent Ohm's law, which is a basic requirement for their function as a nonvolatile memristor.

Ion transporting proteins are known to be memristors [58] and have the distinct advantages of transconductance driven by electrochemical ion gradients, not applied electrical energy, and being naturally nanostructured as compared to electronic memristors. Moreover, ion channels mediate changes in ion fluxes across the membrane to provide the fundamental basis for the membrane depolarizations and hence the generation of action potentials. Since the transconductance is driven by electrochemical ion gradients, not by applied electrical energy as for electronic memristors, the advantage of using ion channels as memristors is that they require 1000-fold less energy for switching compared to state-of-the-art ENODe systems.

The artificial cell described in this second context provides a biological engineering approach to overcome the roadblock of internal heating that limits both the miniaturization and the performance of state-of-the-art electronic devices. Hence, the memristor function of membrane ion-transport proteins is the sufficient basis for controlling membrane information transfer. This is an application of the biological engineering approach described for this artificial cell.

Biological Engineering of Systems to Provide Power

As an example of using ion channels as memristive components, over several years, we have published the electrophysiological results of ion channel function in secretory and excitable cells, which led to the development of an artificial cell with incorporated K^+-permeable ion channels to control membrane transport [6] that could be detected using a Hidden Markhov Chain stochastic model [56]. We used the know-how and biomimetic technology to develop a diffusion device based on the Na^+-dependent membrane transporter NhaA, which we patented as the basis for a Na^+ ion-transporting method of generating electrical energy that utilizes a biomimetic membrane and incorporated ion transport proteins [64].

The research team SyNaBi at the laboratory TIMC in the Université Grenoble Alpes has designed a supported lipid bilayer (SLB) biomimetic membrane system that comprised polyaniline (PANI) to support a lipid bilayer membrane that incorporated NhaA to give the system the capability of controllable electrogenic ion transport (Fig. 2.13). The high turnover rate of NhaA (~ 10^5 per min) provides this

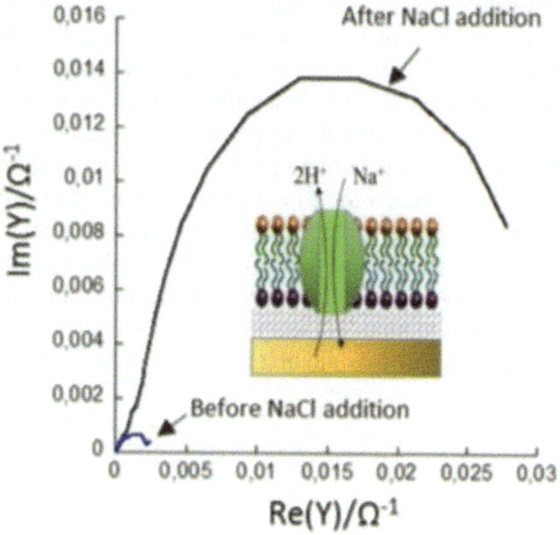

Fig. 2.13 Electrochemical impedance spectroscopy response (Cole–Cole) of SLB-active NhaA-PANI modified gold electrode before and after the addition of NaCl 10 mM. (Reproduced from [66])

PANI-SLB-NhaA system with the function of a high-speed rechargeable biocapacitor, which also provides a low-energy-consuming fast switch for biological engineering applications [65].

The research team SyNaBi, in the laboratory TIMC at the Université Grenoble Alpes, has also developed an implanted biofuel cell (IBFC) that utilizes the redox functions of enzymes to produce electrical energy, particularly when implanted inside the body [66]. We have experienced the development of bioelectrodes that function for long periods when implanted inside the body [67]. Our development of these implanted biofuel cells has provided us with a great deal of know-how in understanding and designing such active autonomous systems for implantation in the body [68].

An implanted biofuel cell (IBFC) is a device that produces power only from the chemicals that are naturally present inside the body, such as glucose and oxygen. It generates electrical power from electrochemical reactions occurring on bioelectrodes. As these sources of energy are always present in the organism, the IBFC can potentially work for infinite time, as long as the substrates can freely diffuse and that the bioelectrodes are not destroyed by the organism. We believe that in the near future, this new generation of green power sources will power implantable electronic devices in order to compensate the failure of function in humans for the rest of their life, from brain stimulators to artificial urinary sphincter. The principle of the IBFC is the oxidation of glucose at the anode by glucose oxidase to supply electrons to the electrical circuit. After passing through the electrical circuit, the electrons are collected at the cathode. At this electrode, the enzyme laccase is used to reduce dioxygen to water. The in vivo proof of concept in a mammal was achieved by Cinquin et al. in 2010 [69]. Several technical problems have been overcome for successfully implanting a full enzymatic biofuel cell. The IBFC has been carefully designed to be miniaturized and made able to operate in the abdominal cavity of a rat. As previously said, since the fuels are not embedded in the IBFC, a very high power/volume ratio is achieved. The bigger the electrodes are, the stronger the power is. This principle helps to miniaturize the electrodes but not the complete device, since the electrodes have been wrapped in several layers of materials in order to provide a biocompatible system. Nonetheless, this biofuel cell supplies enough energy (40 μW) to power electronic devices like a pacemaker with electrodes of 240 μL volume [70]. We recently implanted an enzymatic biofuel cell capable to operate for 2 months in a rabbit [66]. To our knowledge, this is the first long-term implantable device with enzymes implanted with success in animals. Although we have overcome technical problems like miniaturization, sterilization, [71] and monitoring, the issue of sterile inflammatory response and later encapsulation or degradation [67] remains roadblocks to its long-term success as an implanted device. The IBFC is the first implantable device communicating with the body via molecules.

Can a Biofuel Cell Be Defined As a Smart Nanostructured System "Symbio-bot"?

A communicating implant such as an IBFC is required to be a biocompatible system. This leads us to the notion of establishing a nonaggression pact with the body. The IBFC is not only in direct relation with organism's fluids for a long time, but it has to communicate with the body to operate. This is the first kind of implant that exchanges molecules with a behavior of a biochemical redox factory that can output energy or molecules. In a sense, the IBFC functions like an organ, a cell, or an energy factory, similar to mitochondria but much less complex and productive.

The two-way duplex communication is not more an observation but a requirement to provide fuel to the IBFC and enable it to produce electricity, which will indirectly restore the function of the body. Operating in this intimate way, it can be considered as a symbiotic robot. Thus, it has to integrate the body in such a way to establish a stable duplex communication. A long-term integration of this type of device in the body will only succeed if the body accepts at least a "porous encapsulation," preferably transporting the appropriate molecules with a sufficient flow rate.

The development of "symbio-bots" will surely benefit from progress made in the field of medical devices but even more importantly in the field of regenerative medicine. It also paves the way for studies in human and social science. The acceptance of such devices by patients opens a field of interest for philosophy in the context of self-image, for example, with the question of how a "symbio-bot" will be accepted by the patient.

Overcoming Constraints to Allow Nanostructured Systems to Function in the Body

There are many applications for smart nanostructured systems based on biological engineering, both for stand-alone devices and for systems that are designed to function as implanted medical devices. Examples where such systems have advantages for stand-alone devices include ultra-miniaturized light-sensing systems for drone or robotic vision and renewable energy sources. The overarching engineering advantage is the extremely favorable power-to-weight ratio that can be achieved using biological components in a nanostructured system designed for sophisticated biosensing (e.g., vision) or delivery of power. For those two applications, a biological engineered nanostructured system provides the distinct advantages of (i) easy scalability, (ii) easy assembly into 3D structures, and (iii) up to 1000-fold less energy requirements for switching compared to state-of-the-art ENODe systems.

Those are the same advantages for biologically engineered smart nanostructured systems designed to function as implanted medical devices. For an implanted application, the additional, and highly significant, advantage of utilizing biological components is the ease of providing a more natural interface to living tissues. This is because a smart nanostructured system based on biological engineering utilizes biological components that include lipids and proteins. We return to the concept of the "symbio-bot" introduced in the previous section. The "symbio-bot" embodies the advantages of utilizing a biological engineering approach to develop an implanted medical device.

The major challenges to overcome for the function of an implanted "symbio-bot" are the biocompatibility and the interface of the implanted device with the body [68]. In this section, we discuss a roadmap for the design of an implantable "symbio-bot" based on biological engineering. This type of roadmap would be useful, for example, to increase the longevity of bioelectrodes that are designed for implantation in the body [68].

We can consider the flowchart illustrated in Fig. 2.14 as a roadmap of the decision-making themes needed for the development of nanostructured systems that are designed to function in the body. The themes are highlighted by the colored zones in Fig. 2.14 and are summarized in the box below:

Blue zone	In the body *(diagnostic and imaging systems)*
Green zone	Fluid and tissue samples from the body *(diagnostic and imaging systems)*
Purple zone	Appropriate medical therapies *(therapeutic systems and devices)*
Yellow zone	The patient in society *(human, ethical, and societal aspects of diagnostic, imaging, and therapeutic devices)*

Note that a "diagnostic medical device" for diagnostics or imaging can also be created following the stage of regulatory approval and appropriate sterilization that is prior to the input of a doctor for a medical diagnosis. The "diagnostic medical device" at this stage would be one that assists the doctor in achieving an appropriate medical diagnosis. For clarity, a separate symbol to illustrate the "diagnostic medical device" at this stage has been omitted from the flowchart. This flowchart does not emphasize *"technology or materials push"* but instead describes *"medical device pull"* that is based on clinical needs and questions. Nonetheless, such a *"medical device pull"* will need the skills of materials and engineering disciplines to design the "symbio-bot" that is required to solve the clinical needs. The major questions to be answered at the various stages of this roadmap to *"pull"* the medical device are:

1. What is the clinical problem to be solved?
2. Is the diagnostic device functioning inside the body or using fluid or tissues samples from the body (blue and green zones)?
3. Is the diagnostic device easily sterilized, and does it pass regulatory affairs?
4. Does the diagnostic device actually help clinicians with medical diagnosis?
5. Is a diagnostic device needed, or is pharmaceutical therapy preferred (purple zone)?
6. Do patients understand the why the diagnostic device is needed, and is it tolerated by patients (yellow zone)?

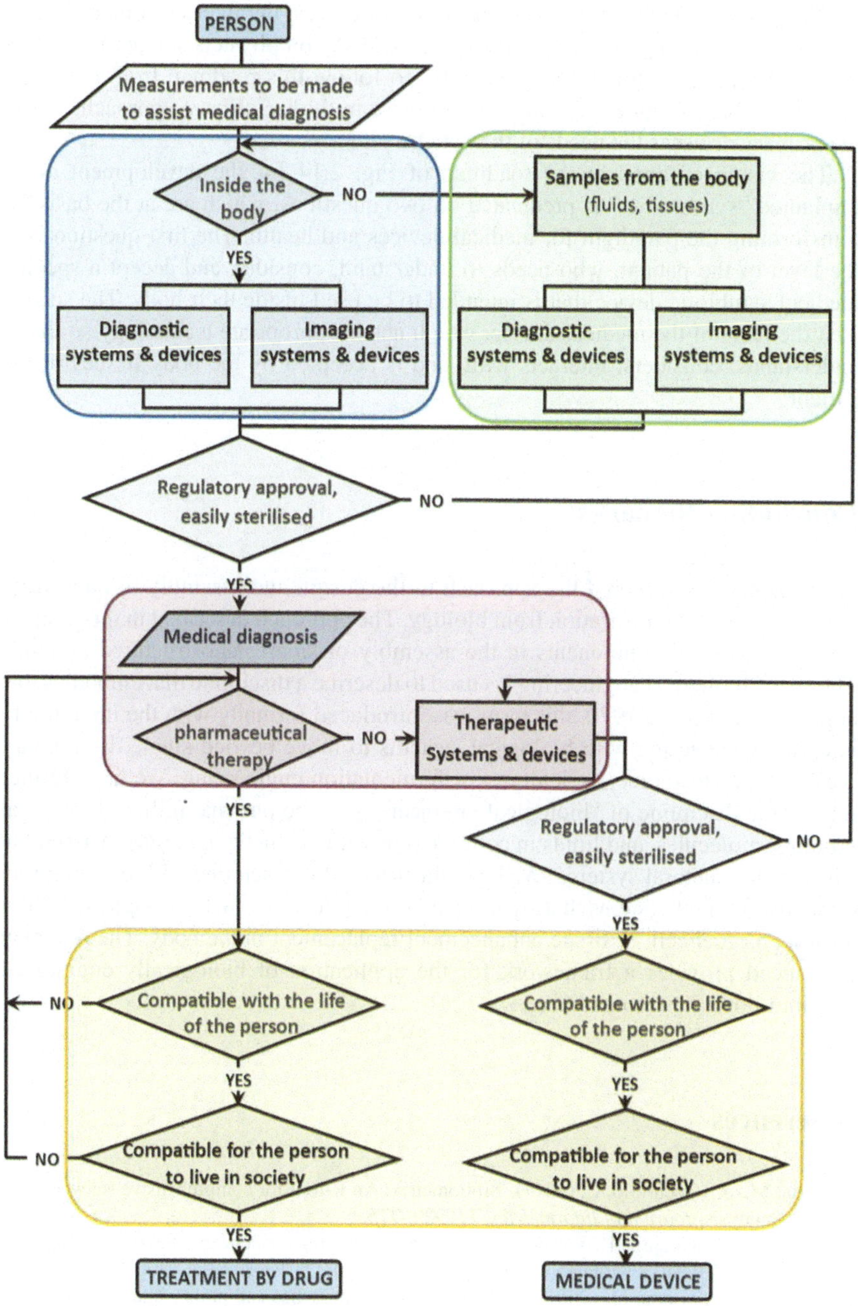

Fig. 2.14 Flowchart of decision-making to develop implantable medical devices

The progression along this roadmap requires at least the interdisciplinary expertise of clinicians, engineers, biologists, biochemists, biophysicists, materials scientists, and regulatory affairs professionals. To follow this roadmap from patient to device or pharmaceutical treatment requires a multidisciplinary approach, where the primary driver is the needs of the patient.

The vision outlined in the roadmap of Fig. 2.14 for the development of an implanted "symbio-bot" is predicated on two questions, which are at the basis for transforming the paradigm for medical devices and health. The first question is at the level of the patient, who needs to understand, consider, and accept a specific medical symbiotic device that is intended to be used inside their body. The second is at the level of the medical device, which needs appropriate technology so that it understands, considers, interacts with, and is accepted by the body tissues of the patient.

Concluding Remarks

This chapter has explored the approach to the design and assembly of nanostructured systems with inspiration from biology. The approach described in this chapter utilizes biological components in the assembly of smart nanostructured systems. The term "biological engineering" is used to describe a discipline that embodies this approach. Indeed, in 1970, this term was introduced formally with the intention to integrate engineering with biological systems to move beyond single disciplinary areas such as medicine, agriculture, or fermentation engineering. We have further refined the discipline of "biological engineering" to be one that utilizes biological proteins, molecules, and lipids in combination with synthetic materials to assemble smart nanostructural systems. We have illustrated this discipline with the assembly of nanostructured systems that are targeted for applications such as diagnostic, therapeutic, biofuel cell, or tissue enhancement/replacement in the body. The roadmap introduced provides a framework for the application of biologically engineered "symbio-bots" in nanomedicine.

References

1. Aria, M., & Cuccurullo, C. (2017). Bibliometrix: An R-tool for comprehensive science mapping analysis. *Journal of Informetrics, 11*, 929–975.
2. Alcaraz, J. P., Cinquin, P., & Martin, D. K. (2018). Tackling the concept of symbiotic implantable medical devices with nanotechnologies. *Biotechnology Journal, 13*, 1800102.
3. Hall, S. G., & Lima, M. (2001). Problem-solving approaches and philosophies in biological engineering: Challenges form technical, social, and ethical arenas. *Transaction of the ASAE, 44*, 1037–1041.
4. Johnson, A. T. (2006). The making of a new discipline. *International Journal of Engineering Education, 22*, 3–8.

5. Tao, B. Y., Allen, D. K., & Okos, M. R. (2006). The evolution of biological engineering. *International Journal of Engineering Education, 22*, 45–52.
6. Battle, A. R., Valenzuela, S. M., Mechler, A., Nichols, R. J., Praporski, S., di Maio, I. L., Islam, H., Girard-Egrot, A. P., Cornell, B. A., Prashar, J., Caruso, F., Martin, L. L., & Martin, D. K. (2009). Novel engineered ion channel provides controllable ion permeability for polyelectrolyte microcapsules coated with a lipid membrane. *Advanced Functional Materials, 19*, 201–208.
7. Rini, B. I., Campbell, S. C., & Escudier, B. (2009). Renal cell carcinoma. *Lancet, 373*, 1119–1132.
8. Zou, Y., Palte, M. J., Deik, A. A., Li, H., Eaton, J. K., Wang, W., et al. (2019). A GPX4-dependent cancer cell state underlies the clear-cell morphology and confers sensitivity to ferroptosis. *Nature Communications, 10*, 1617.
9. Uhlén, M., Fagerberg, L., Hallström, B. M., Lindskog, C., Oksvold, P., Mardinoglu, A., et al. (2015). Tissue-based map of the human proteome. *Science, 347*, 1260419.
10. Bolla, J. R., Agasid, M. T., Mehmood, S., & Robinson, C. V. (2019). Membrane protein-lipid interactions probed using mass spectrometry. *Annual Review of Biochemistry, 88*, 85–111.
11. Rask-Andersen, M., Almén, M. S., & Schiöth, H. B. (2011). Trends in the exploitation of novel drug targets. *Nature Reviews Drug Discovery, 10*, 579–590.
12. Liu, Z., Liu, Y., Xu, L., An, H., Chang, Y., Yang, Y., Zhang, W., & Xu, J. (2015). P2X7 receptor predicts postoperative cancer-specific survival of patients with clear-cell renal cell carcinoma. *Cancer Science, 106*, 1224–1231.
13. Burnstock, G., & Knight, G. E. (2018). The potential of P2X7 receptors as a therapeutic target, including inflammation and tumour progression. *Purinergic Signalling, 14*, 1–8.
14. Yanda, M. K., Liu, Q., & Cebotaru, L. (2018). A potential strategy for reducing cysts in autosomal dominant polycystic kidney disease with a CFTR corrector. *Journal of Biological Chemistry, 293*, 11513–11526.
15. Carson, C., Raman, P., Tullai, J., Xu, L., Henault, M., Thomas, E., Yeola, S., Lao, J., McPate, M., Verkuyl, J. M., Marsh, G., Sarber, J., Amari, A., Bailey, S., Lubicka, D., Pham, H., Miranda, N., Ding, J., Tang, H. M., Ju, H., Tranter, P., Ji, N., Krastel, P., Jain, R. K., Schumacher, A. M., Loureiro, J. J., Goerge, E., Berellini, G., Ross, N. T., Bushell, S. M., Erdemli, G., & Solomon, J. M. (2015). Englerin A agonizes the TRPC4/C5 cation channel to inhibit tumor cell line proliferation. *PLos ONE, 10*, e0127498.
16. Mijatovic, T., Dufraasne, F., & Kiss, R. (2012). Cardiotonic steroids-mediating targeting of the Na(+)/K(+)-ATPase to combat chemoresistant cancers. *Current Medicinal Chemistry, 19*, 627–646.
17. Srinivasan, B., Kolli, A. R., Esch, M. B., Abaci, H. E., Shuler, M. L., & Hickman, J. J. (2015). TEER measurement tchniques for in vitro barrier model systems. *Journal of Laboratory Automation, 20*, 107–216.
18. Tang, V. W., & Goodenough, D. A. (2003). Paracellular ion channel at the tight junction. *Biophysical Journal, 84*, 1660–1673.
19. Aung, P. P., Mitani, Y., Sanada, Y., Nakayama, H., Matsusaki, K., & Yasui, W. (2006). Differential expression of claudin-2 in normal tissues and gastrointestinal carcinomas. *Virchows Archiv, 448*, 428–434.
20. Singh, A. B., & Harris, R. C. (2004). Epidermal growth factor receptor activation differentially regulates claudin expression and enhances transepithelial resistance in Madin-Darby Canine Kidney Cells. *Journal of Biological Chemistry, 279*, 3543–3552.
21. Cereijido, M., Contreras, R. G., Flores-Benitez, D., Flores-Maldonado, C., Larre, I., Ruiz, A., & Shoshani, L. (2007). New diseases derived or associated with the tight junction. *Archives of Medical Research, 38*, 465–478.
22. Flores-Benitez, D., Ruiz-Cabrera, A., Flores-Maldonado, C., Shoshani, L., Cereijido, M., & Contreras, R. G. (2007). Control of tight junctional sealing: role of epidermal growth factor. *American Journal of Physiology Renal Physiology, 292*, F828–F836.

23. Flores-Benitez, D., Rincon-Heredia, R., Razgado, L. F., Larre, I., Cereijido, M., & Contreras, R. G. (2009). Control of tight junctional sealing: roles of epidermal growth factor and prostaglandin E_2. *American Journal of Physiology Cell Physiology, 297*, C611–C620.
24. Tanos, B. E., Perez Bay, A. E., Salvarezza, S., Vivanco, I., Mellinghoff, I., Osman, M., Sacks, D. B., & Rodriguez-Boulan, E. (2015). IQGAP1 controls tight junction formation through differential regulation of claudin recruitment. *Journal of Cell Science, 128*, 853–862.
25. Liu, G. J., Simpson, A. M., Swan, A. M., Tuch, B. E., & Martin, D. K. (2003). ATP-sensitive potassium channels induced in liver cells after transfection with insulin receptor and GLUT2 transporter regulate glucose-stimulated insulin secretion. *FASEB Journal, 17*, 1682–1684.
26. Eisenbarth, G. S. (1986). Type I diabetes mellitus: a chronic autoimmune disease. *New England Journal of Medicine, 14*, 1360–1368.
27. The Diabetes Control and Complications Trial Research Group. (1993). The effect of intensive treatment of diabetes on the development and progression of long-term complications in insulin-dependent diabetes mellitus. *New England Journal of Medicine, 329*, 977–986.
28. Cook, D. L., & Hales, C. N. (1984). Intracellular ATP directly blocks K+ channels in pancreatic β cells. *Nature, 311*, 271–273.
29. Ashcroft, F. M., Harrison, D. E., & Ashcroft, S. J. H. (1984). Glucose induces closure of single potassium channels in isolated rat pancreatic β cells. *Nature, 312*, 446–468.
30. Pralong, W. K., Bartley, C., & Wollheim, C. B. (1990). Single islet β cell stimulation by nutrients: relationship between pyridine nucleotides, cytosolic Ca2+ and secretion. *EMBO Journal, 9*, 53–60.
31. Sakurada, M., Kanasuka, A., Saitoh, T., Makino, H., Yamamura, K., Miyazaki, J., Kikuchi, M., & Yoshida, S. (1993). Relation between glucose-stimulated insulin secretion and intracellular calcium accumulation studies with a super fusion system of a glucoseresponsive pancreatic β cell line MIN6. *Endocrinology, 132*, 2659–2665.
32. Aguilar-Bryan, L., Nichols, C. G., Wechsler, S. W., Clement, J. P., Boyd, A. E., Gonzalez, G., Herrera-Sosa, H., Nguy, K., Bryan, J., & Nelson, D. A. (1995). Cloning of the β cell high affinity sulfonylurea receptor: a regulator of insulin secretion. *Science, 268*, 423–426.
33. Inagaki, N., Gonoi, T., Clement, J. P., Namba, N., Inazawa, J., Gonzalez, G., Aguilar-Bryan, L., Seino, S., & Bryan, J. (1995). Reconstitution of I_{KATP}: an inward rectifier subunit plus the sulfonylurea receptor. *Science, 270*, 1166–1170.
34. Sakura, H., Ammala, C., Smith, P. A., Gribble, F. M., & Ashcroft, F. M. (1995). Cloning and functional expression of the cDNA encoding a novel ATP-sensitive potassium channel expressed in pancreatic β cells, brain, heart and skeletal muscle. *FEBS Letters, 377*, 338–344.
35. Babenko, A. P., Aguilar-Bryan, L., & Bryan, J. (1998). A view of sur/KIR6.X K_{ATP} channels. *Annual Review of Physiology, 60*, 308–315.
36. Spruce, A. E., Standen, N. B., & Stanfield, P. R. (1985). Voltage-dependent ATP-sensitive potassium channels of skeletal muscle membrane. *Nature, 316*, 736–738.
37. Noma, A. (1983). ATP-regulated K+ channels in cardiac muscle. *Nature, 305*, 147–148.
38. Wellman, G. C., Barrett-Jolley, R., Koppel, H., Everitt, D., & Quayle, J. M. (1999). Inhibition of vascular K(ATP) channels by U-37883A: a comparison with cardiac and skeletal muscle. *British Journal of Pharmacology, 128*, 909–916.
39. Lee, K., Dixon, A. K., Freeman, T. C., & Richardson, P. J. (1998). Identification of an ATP-sensitive potassium channel current in rat striatal cholinergic interneurones. *Journal of Physiology, 510*, 441–453.
40. Russ, U., Rauch, U., & Quast, U. (1999). Pharmacological evidence for a KATP channel in renin-secreting cells from rat kidney. *Journal of Physiology, 517*, 781–790.
41. Henderson, R. M., Graf, J., & Boyer, J. L. (1989). Inward-rectifying potassium channels in rat hepatocytes. *American Journal of Physiology, 171*, G1028–G1035.
42. Capiod, T., & Ogden, D. C. (1989). The properties of calcium-activated potassium ion channels in Guinea-pig isolated hepatocytes. *Journal of Physiology, 409*, 285–295.
43. Lidofsky, S. D. (1995). Convergent and parallel activation of low-conductance potassium channels by calcium and cAMP-dependent protein kinase. *PNAS, 92*, 7115–7119.

44. Simpson, A. M., Tuch, B. E., Swan, M. A., Tu, J., & Marshall, G. M. (1995). Functional expression of the human insulin gene in a human hepatoma cell line (HEPG2). *Gene Therapy, 2*, 223–231.
45. Simpson, A. M., Marshall, G. M., Tuch, B. E., Maxwell, L., Szymanska, B., Tu, J., Beynon, S., Swan, M. A., & Camacho, M. (1997). Gene therapy of diabetes: glucose stimulated insulin secretion in a human hepatoma cell line (HEPG2ins/g). *Gene Therapy, 4*, 1202–1215.
46. Simpson, A. M., Szymanska, B., Tuch, B. E., & Marshall, G. M. (1999). Secretion and storage of insulin from a human hepatoma cell line: Huh7-ins. *Transplantation Proceedings, 31*, 812.
47. Ferber, S., Halkin, A., Cohen, H., Ber, I., Einav, Y., Goldberg, I., Barshack, I., Seijffers, R., Kopolovic, J., Kaiser, N., & Karasik, A. (2000). Pancreatic and duodenal homeobox gene I induces expression of insulin genes in liver and ameliorates streptozotocin-induced hyperglycaemia. *Nature Medicine, 6*, 568–572.
48. Soria, B., Roche, E., Berna, G., Leon-Quinto, T., Reig, J. A., & Martin, F. (2000). Insulin secreting cells derived from embryonic stem cells normalize glycaemia in streptozotocin induced diabetic mice. *Diabetes, 49*, 157–162.
49. Tuch, B. E., Szymanska, B., Yao, M., Tabiin, M. T., Gross, D. J., Holman, S., Swan, M. A., Humphrey, R. K. B., Marshall, G. M., & Simpson, A. M. (2003). Function of a genetically modified human liver cell line that stores, processes and secretes insulin. *Gene Therapy, 10*, 490–503.
50. Permutt, M. A., Koranyi, L., Keller, K., Lacy, P. E., Scharp, D. M., & Mueckler, M. (1989). Cloning and functional expression of a human pancreatic islet glucose-transporter cDNA. *PNAS, 86*, 8688–8692.
51. Weinhouse, S. (1976). Regulation of glucokinase in liver. In B. L. Horecker & E. R. Stadtman (Eds.), *Current Topics in Cellular Regulation* (pp. 1–50). Academic Press.
52. Liu, G. J., Simpson, A. M., Swan, M. A., Tao, C., Tuch, B. E., Crawford, R. M., Jovanovic, A., & Martin, D. K. (2003). ATP-sensitive potassium channels induced in liver cells after transfection with insulin cDNA and the GLUT2 transporter regulate glucoe-stimulated insulin secretion. *FASEB Journal, 17*, 1682–1684.
53. Liu, G. J., Simpson, A. M., Swan, M. A., Tao, C., Tuch, B. E., Crawford, R. M., Jovanovic, A., & Martin, D. K. (2003). ATP-sensitive potassium channels induced in liver cells after transfection with insulin cDNA and the GLUT2 transporter regulate glucoe-stimulated insulin secretion. *FASEB Journal Express, 18*. https://doi.org/10.1096/fj.02-0051fje
54. Henquin, J. C. (2000). Triggering and amplifying pathways of regulation of insulin secretion by glucose. *Diabetes, 49*, 1751–1760.
55. Lindemann, M., & Winterhalter, M. (2006). Membrane channels as a tool to control nanoreactors. *IEE Proceedings Nanobiotechnology, 153*, 107–111.
56. Krishnamurthy, V., Luk, K. Y., Cornell, B., Prashar, J., di Maio, I. L., Islam, H., Battle, A., Valenzuela, S., & Martin, D. K. (2007). Gramicidin ion channel based nano-biosensors: Construction, stochastic dynamical models and statistical detection algorithms. *IEEE Sensors Journal, 7*, 1281–1288.
57. Köhler, G., Moya, S. E., Leporatti, S., Bitterlich, C., & Donath, E. (2007). Stability and fusion of lipid bilayers on polyelectrolyte multilayer supports studies by colloidal force spectroscopy. *European Biophysics Journal, 36*, 337–347.
58. Chua, L. O. (1971). Memristor—the missing circuit element. *IEEE Transactions on Circuit Theory, CT-18*, 507–519.
59. Strukov, D. B., Snider, G. S., Stewart, D. R., & Williams, R. S. (2008). The missing memristor found. *Nature, 453*, 80–83.
60. Ziegler, M., Riggert, C., Hansen, M., Bartsch, T., & Kohlstedt, H. (2015). Memristive Hebbian plasticity model: device requirements for the emulation of Hebbian plasticity based on memristive devices. *IEEE Transactions on Biomedical Circuits and Systems, 9*, 197–206.
61. Li, Y., Zhong, Y., Zhang, J., Xu, L., Wang, Q., Sun, H., Tong, H., Cheng, X., & Miao, X. (2014). Activity-dependent synaptic plasticity of a chalcogenide electronic synapse for neuromorphic systems. *Scientific Reports, 4*, 4906.

62. Saïghi, S., Mayr, C. G., Serrano-Gotarredona, T., Schmidt, H., Lecerf, G., Tomas, J., Grollier, J., Boyn, S., Vincent, A. F., Querlioz, D., La Barbera, S., Alibart, F., Vuillaume, D., Bichler, O., Gamrat, C., & Linares-Barranco, B. (2015). Plasticity in memristive devices for spiking neural networks. *Frontiers of Neuroscience, 9*, 51.
63. Boybat, I., Le Gallo, M., Nandakumar, S. R., Moraitis, T., Parnell, T., Tuma, T., Rajendran, B., Leblebici, Y., Sebastian, A., & Eleftheriou, E. (2018). Neuromorphic computing with multi-memristive synapses. *Nature Commununications, 9*, 25.
64. Cinquin, P., & Martin, D. K. (2017). Biomimetic artificial membrane device. *US, 9*(577), 280.
65. Ben Tahar, A., Zebda, A., Alcaraz, J. P., Gayet, L., Boualam, A., Cinquin, P., & Martin, D. K. (2019). A PANI supported lipid bilayer that contains NhaA transporter proteins provides a basis for a biomimetic biocapacitor. *Chemical Communications, 55*, 13152–13155.
66. El Ichi-Ribault, S., Alcaraz, J. P., Boucher, F., Boutaud, B., Dalmolin, R., Boutonnat, J., Zebda, A., Cinquin, P., & Martin, D. K. (2018). Remote wireless control of an enzymatic biofuel cell implanted in a rabbit for 2 months. *Electrochimica Acta, 269*, 360–366.
67. El Ichi, S., Zebda, A., Alcaraz, J. P., Laaroussi, A., Boucher, F., Boutonnat, J., Reverdy-Bruas, N., Chaussy, D., Belgacem, M. N., Cinquin, P., & Martin, D. K. (2015). Bioelectrodes modified with chitosan for long-term energy supply from the body. *Energy Environmental Science, 8*, 1017–1026.
68. Zebda, A., Alcaraz, J. P., Vadgama, P., Shleev, S., Minter, S., Boucher, F., Cinquin, P., & Martin, D. K. (2018). Challenges for successful implantation of biofuel cells. *Bioelectrochemistry, 124*, 57–72.
69. Cinquin, P., Gondran, C., Giroud, F., Mazabrard, S., Pellissier, A., Boucher, F., Alcaraz, J. P., Gorgy, K., Lenouvel, F., Mathe, S., Porcu, P., & Cosnier, S. (2010). A glucose biofuel cell implanted in rats. *PLoS One, 5*, e10476.
70. Zebda, A., Cosnier, S., Alcaraz, J. P., Holzinger, M., Le Goff, A., Gondran, C., Boucher, F., Giroud, F., Gorgy, K., Lamraoui, H., & Cinquin, P. (2013). Single glucose biofuel cells implanted in rats power electronic devices. *Scientific Reports, 3*, 1516.
71. Alcaraz, J. P., El Ichi-Ribault, S., Cortella, L., Guimier-Pingault, C., Zebda, A., Cinquin, P., & Martin, D. K. (2016). La biopile enzymatique à glucose/oxygène: Quelques nuances de Grays. *Médecine/Sciences, 32*, 771–773.

Chapter 3
Role for Biological Engineering of "Symbio-Bots" in Nanomedicine

In this chapter, we will explore the practical aspects of biomimetic membrane technology in nanomedicine. This focus on practical aspects will include a description of a refined definition of the discipline of biological engineering. This refined discipline of biological engineering includes the aspect of human factors concurrently with technical factors, particularly when we consider the therapeutic and diagnostic needs for nanomedicine. Such needs rely upon medical devices, either implanted or in contact with the body of a patient. The enhancements to such medical devices will require them to be "smart," in the sense that there will need to be an improved understanding and construction of such devices to provide for long-term two-way (duplex) communication of biological molecules and signals to and from the device to the body. Concurrently, the patient needs to be accepting of such innovative "smart" devices that will have the capability to function autonomously. That is, the patient needs to understand and provide permission for the "smart" device to function autonomously. Here we return to the roadmap presented in Chap. 2 (Fig. 2.14) and the six questions that are at the heart of the *"medical device pull"* to utilize skills of materials and engineering disciplines to design the "symbio-bot" that is required to solve the clinical needs. In this chapter, major questions to be answered at the various stages of this roadmap to *"pull"* the medical device are:

1. What is the clinical problem to be solved?
2. Is the diagnostic device functioning inside the body or using fluid or tissue samples from the body (blue and green zones)?
3. Is the diagnostic device easily sterilized, and does it pass regulatory affairs?
4. Does the diagnostic device actually help clinicians with medical diagnosis?
5. Is a diagnostic device needed, or is pharmaceutical therapy preferred (purple zone)?
6. Do patients understand why the diagnostic device is needed, and is it tolerated by patients (yellow zone)?

We will first discuss this enhanced duplex communication in the context of symbiotic medical devices, which will need to incorporate biomimetic membrane

D. Martin, I. Vilgrain, *Biomimetic Membrane Technology*,
https://doi.org/10.1007/978-3-031-90594-0_3

technology so that long-term duplex communication can occur unhindered by inflammatory processes that currently lead to rejection or encapsulation of an implanted medical device.

Why Does This "Symbio-Bot" Vision Propose a Transformation of the Paradigm for Medical Devices and Health?

In the current research environment of nanobiotechnologies, the proposed paradigm shift for a "symbio-bot" provides the opportunity to exploit the advances achieved already with symbiotic medical devices but within a framework of human factors. The "symbio-bot" will have biomimetic membrane interfaces to provide a two-way communication of molecules with the tissues of the body. That is, the "symbio-bot" is provided with an unprecedented technical level of authority to communicate with the tissues of the body. The human aspect of delegating such technical authority requires that the patient understand and permit the autonomous function of such a symbiotic medical device with their body. One of the best-known examples of such acceptance is the implantation of pacemakers, which requires a very high level of confidence in the technology and a large positive benefit/risk balance.

There are three overarching issues that provide a framework for this paradigm shift to develop a "symbio-bot":

Issue(1) The need to enhance the psychosocial representations/perceptions of symbiotic medical devices ("symbio-bot") and to develop tools to enhance targeted patients' practical understanding of these novel "symbio-bot."

Issue(2) The need to enhance the interface between "symbio-bot" and the body to enable stable biocompatible function during long-term implantation. This needs to be achieved by defining a detailed molecular-level understanding of the nanostructure and function of "symbio-bot," and to utilizing "smart" matrices with integrated biosensors for the materials used to assemble a "symbio-bot."

Issue(3) The need to develop the discipline of biological engineering with biological proteins, synthetic materials, and biomimetic membranes to create "symbio-bot" for long-term implantation.

Structuring the Paradigm Shift for the Development of "Symbio-Bots"

This multidisciplinary approach required for a "symbio-bot" necessitates a shift away from the traditional partitioning of scientific disciplines into linear subsets for the development of medical devices. In the classic linear development progression for a medical device, there are five phases that move from planning, design, prototype, design for manufacturing, and manufacturing. This linear progression does not easily provide the necessary and constant feedback with end users (i.e., medical doctors and patients), which results in inappropriate medical devices that require redesign and adjustments, if these are indeed possible. The multidisciplinary approach of "symbio-bot" easily provides a circular production process to ensure the end-users interact with the different scientific disciplines in all production phases.

Managing the emergence of biological engineering to optimize its multidisciplinary contribution to outcomes in nanomedicine requires a nontraditional approach. Otherwise, the growth of nanomedicine will remain constrained by the dichotomy of whether the starting point for advances in nanobiotechnology and nanomedicine is in the realm of fundamental scientists in traditional disciplines to first seek knowledge in the basic functions of molecules that comprise biomimetic membranes or in the realm of clinicians and engineers who first seek to observe and characterize defined medical problems and needs. A biological engineering approach, which includes biomimetic membrane technology, commences with the premise that the value of nanomedicine is to improve the health of individuals. A related premise is that the value of biomimetic membrane technology provides a symbiotic interface to improve the function of systems and devices that interact with tissues and cells of the body. Indeed, such a biological engineering approach allows contributions from both a fundamental and an applied approach within the same project.

Positioning "Symbio-Bots" Within an Industrial, Healthcare, and Medical Device Framework

The multidisciplinary approach to the development of "symbio-bots" provides the shift in paradigm to efficiently deliver healthcare in the context of these redefinitions of healthcare. Current approaches for health innovations are mostly based on a techno-determinist paradigm. Even if some user-centered approaches exist, introducing human motivations, perceptions, representations, and practices into innovation uses in health is not sufficient to develop new devices full potential and prevent its unintended consequences. A focus on transformations of body representations will be done in the context of modern societies dominated by the cybernetic model. About the specificities of these new symbiotic systems of control and communication, the paradigm of the post-human body as a "machine" will be completed and debated. This new paradigm of representations of the body and its capabilities needs to be modeled including maps of the transformations of mental frontiers between natural and artificial, normal and deviant, power and weakness, etc.

The main challenge of codesign with targeted patients is to identify and characterize all these transformations and to address the associated ethical and social issues. Although end-user-centered approaches have included "human in the loop," the key ethical and social issues related to the use of "symbio-bots" in health settings have not been clearly considered. These ethical and social issues include the autonomy of the device, delegation of authority and responsibility, social connectedness, health literacy, patient engagement, and integration in daily self-care practices.

Healthcare sustains a major revolution to evolve from a reactive management (dealing with the management of an established pathology validated by an ad hoc medical diagnosis) to preventive management. According to the WHO, prevention can be defined as all measures that are aimed at avoiding or reducing the number and severity of diseases, accidents, and disabilities through reliable diagnostics, monitoring, and or treatment. Prevention can be divided into three areas dealing with *(i)* early detection and treatment of a disease, *(ii)* the actions to reduce the progression and complications of a proven disease or health problem, and *(iii)* the actions related to medical and rehabilitative care.

External factors such as growing and ageing population, along with a greater prevalence of chronic diseases and the emergence of medical deserts, require rethinking the medical habits. These societal evolutions, as pointed out by the European Union [1], require healthcare solutions that are coordinated, convenient (able to adapt to tissue conformation and movement without disconnections and/or tissue injury), ensure accurate and real time measurements, enable delocalized patient monitoring/treatment, and are accessible at an affordable price [2]. Among them, wearable and implanted medical devices are of primary interest to address these requirements.

At an industrial level, the conjunction of the aforementioned desired evolutions and undergone constraints has strongly reinforced the need for new medical devices, leading to a continuously increasing healthcare market (+2.9% in 2013–2017, +5.4% forecasted for 2018–2022). By itself, the wearable healthcare market is expected to be worth $US46.6 billion by 2025. If numerous delocalized (self)diagnostics tools have already entered the healthcare market with blockbuster systems such as glucose biosensors for diabetes management or I-Stat multiparametric analyzer for in-field emergency physiological diagnostics, wearable and implanted connected medical devices dedicated to real-time patient monitoring and/or treatment are less developed. More particularly, the emerging sector of smart, hybrid, and conformable medical devices constitutes a great market opportunity and differentiator. It can boost the French med-tech industry, which is suffering from the competition with low-cost production countries (India, China) but also with countries that have succeeded in establishing strategic, technological, and economic transformations (USA, Switzerland, Germany, Italy). In particular the French very dynamic pool of start-ups and SMEs from both the health/med-tech and the electronics sectors could play a decisive role in the development of these future MDs. The medical devices sector is characterized by a very strong dominance of SMEs (92% of companies in France), reflecting a rich and highly diversified industrial network. The 1500 companies, including nearly 200 start-ups, registered on national territory in 2019 have a turnover of 30 billion euros and represent 90,000 jobs. The SNITEM analysis conducted in 2019 confirms the dynamism of the sector with a growth of +200 companies over the last 2 years and an export turnover (9 billion euros) growing strongly by 10% per year [3]. The sector is highly innovative, with 3750 patents filed per year and two-thirds of companies engaged in R&D programs. However, this sector suffers from the lack of dedicated equipment and specific know-how to support rapid prototyping at an affordable cost.

The significant opportunity is to frame a new health paradigm, which will provide exceptional material to challenge all the scientific disciplines, including those from humanities and social sciences. This is particularly important in constructing the philosophical and ethical framework for this e-health, developing expertise in user experience design, including codesign with patients, and in corporal/emotional communication/design. The explicit response for position of "symbio-bot" is the process of understanding and appropriating this new paradigm for patients, to then demonstrate the relevance of cross-disciplinarity at theoretical and methodological complementarity.

The challenge is to demonstrate whether technologies such as "symbio-bots" transform habits and expectations, traditionally built around the natural limits of the human being and body that are intuitively obvious to people. Such limits are also shifted so that e-health and connected health take on a meaning with regard to a realization of the model of the vision of medicine as predictive, preventive, personalized, and participatory ("P4 medicine").

Of important relevance to this vision of "symbio-bots" to transform the paradigm for medical devices and health, in the context of the healthcare sector, Deloitte [2] has commented that:

> Is the long-awaited seismic shift in health care finally here? A collision of forces—a global pandemic of historic proportions; exponential advances in medical science; an explosion of digital technologies, data access, and analytics; informed and empowered consumers; and a movement from disease care to prevention and well-being—proving to be the catalyst for the clinical, financial, and operational transformation that health care has long promised to the world.

The Deloitte report [2] defines the major issues confronting the healthcare sector as:

- *Health Equity:* The ability to fulfill our human potential in all aspects of health and well-being. It's an opportunity to achieve an overall state of well-being encompassing clinical, mental, social, emotional, physical, and spiritual health, and it is influenced by not just healthcare, but also social, economic, and environmental factors.
- *Environmental, Social and Governance (ESG):* The public health community has dubbed climate change the twenty-first century's greatest threat to public health.
- *Mental Health and Well-Being Health:* Recent health, climate, and political crises have both highlighted and exacerbated the world's mental health challenges, which are moving up the priority list in government, provider, health plan, and employer health equity discussions.

- *Digital Transformation and Healthcare Delivery Model Convergence:* Travel bans are making it difficult to recruit foreign staff, so governments have been enlisting retired doctors and nurses and medical students to assist with patient care. One solution lies in digital transformation and healthcare delivery model (HCDM) convergence, which is a trend that has accelerated during the COVID-19 pandemic.
- *Future of Medical Science:* Medical science is currently being transformed by scientific discoveries that will dramatically advance the way we diagnose and treat different diseases. Breakthroughs in digital medicines, nanomedicine, genomics, microbiometrics/metabolomics, and others are occurring at an unprecedented and exponential pace, building on transformed clinical trial structures and timing, disruptive market entrants, and increased consumer participation.
- *Public Health Reimagined:* The pandemic has ignited growing recognition of the need to invest in population health. We are only as strong as our most vulnerable populations, and early health promotion, detection, and intervention are essential to prevent, reduce, or delay the onset of chronic disease.

The roadmap illustrated in Chap. 2 (Fig. 2.14) provides a vision for a multidisciplinary research program that addresses directly the issue of "Future of Medical Science." We can construct a program of research and development at two levels. The first "patient-level" directly implicates human and societal needs and wants in the research and innovation for symbiotic medical devices. The second "device-level" directly implicates the needs to improve its interface with the tissues of the body. The roadmap includes human factors as a concurrent central theme alongside the technical research that is required to develop medical devices. This concurrence aims to overcome the perennial problem of existing medical device developments, which is the reluctance of many groups of people that don't want to engage with things of a scientific basis, including medicines and treatments. The COVID-19 vaccine hesitancy is a very recent example. The roadmap paradigm of concurrent codesign with human and technical factors provides a significant step forward to overcome the reluctance of healthcare providers, insurers, and patients to engage with new medical interventions and devices, even after they have been cleared by regulatory authorities. Consequently, the roadmap also addresses the societal issues of "Health Equity" and "Mental Health and Well-Being Health."

An important consequence of the "symbio-bot" transformation of the medical device and health paradigm is that it naturally includes the ideals of Responsible Research Innovation (RRI) in its research program [4]. The most common definition of RRI was provided by von Schomberg [5] as:

> RRI is a transparent, interactive process by which societal actors and innovators become mutually responsive to each other with a view to the (ethical) acceptability, sustainability and societal desirability of the innovation process and its marketable products (in order to allow a proper embedding of scientific and technological advances in our society).

In the context of the roadmap to illustrate the development of "symbio-bots" responding directly to societal issues, it is important to consider that this roadmap is not simply "technology-push" but has a multidisciplinary research program that embeds the concerns of "societal needs and wants" directly into the codesign processes for novel symbiotic medical devices. Thus, research and innovation tasks that lead to the development of "symbio-bots" go beyond simply being driven by the Technology Readiness Level (TRL) but concurrently include the needs for the Societal Readiness Level (SRL).

Managing Multidisciplinarity to Achieve a Successful "Symbio-Bot"

At this point we can revisit the concluding points at the end of Chap. 1, where we introduced the need for a paradigm shift. We can consider a context of duplex communication at the patient level (SRL) and at the technical device level (TRL). The vision of the paradigm shift is to improve the interface between targeted patients and a medical device.

We will illustrate our vision for a paradigm shift by introducing a practical example. The practical example we will use is a "symbio-bot" that we can define as a medical device that is covered by a smart porous membrane to provide an interface that allows a duplex communication of molecules with the body [6]. Duplex communication means that molecules move in the both directions from the device to the body and from the body to the device. To achieve this type of duplex communication to be stable over time of implantation, the smart porous membrane for a "symbio-bot" is not simply some form of size-exclusion membrane, but rather it mimics closely the selective transport properties of the cell membrane. Such a membrane would be assembled to maintain duplex communication using biomimetic membrane technology. Some examples of "symbio-bots" where this type of biomimetic membrane is necessary to maintain duplex communication include an implantable enzymatic biofuel cell that generates power from the glucose present in the extracellular fluid [7], a biomimetic artificial membrane device that generates power by transporting Na^+ ions from the extracellular fluid [8], an ingestible polymeric capsule that samples fluids from the gastrointestinal tract [9], an implantable polymeric membrane device that modulates the microbiome of the gastrointestinal system for medical therapy [10], a 3D cell culture lab-on-chip microfluidic device for real-time liquid biopsy [11, 12], and smart matrices based on cross-linked collagen hydrogels with bio-optical sensors for online monitoring of $pO_{2matrix}$, pH_{matrix} in cell therapy applications [13].

The paradigm shift has two levels, which are at the interface of the "symbio-bot" with body tissues so it can maintain duplex symbiotic communication with the body, and secondly at the interface of the "symbio-bot" with the patient who needs to accept and trust that these "symbio-bots" will function symbiotically, and likely autonomously, in their body.

The specific technical expertise required to develop the interface of the "symbio-bot" with body tissues includes nano- and micro-biotechnology, biomimetic membranes, polymers, biomaterials, bioelectrochemistry, electrophysiology, biophysics, bioengineering, and medical systems design. Moreover, to achieve the interface of the "symbio-bot" with the patient requires extensive expertise in human factors. The successful "symbio-bot" that maintains both interfaces will need the combined technical and human factor expertise to achieve a multidisciplinary performance. This is not simply an interdisciplinary side-by-side amalgmation of expertise, but is a deliberately managed multidisciplinary design and development process from the start. This presents a major challenge for the existing interdisciplinary concepts of aligning structures of the required expertise from the technical and human factor disciplines in a side-by-side manner. Hence, the need to take a fresh approach and the paradigm shift in the design and development of symbiotic medical device "symbio-bots."

Serious Gaming and User Experience as Important Tools for Framing the Paradigm Shift

Clearly, the challenge is how to manage multidisciplinarity from the start of the design and development of a "symbio-bot." An approach to this challenge is to adopt the techniques of the user experience (UX) that are utilized by social and humanities disciplines. Such a UX is an important aspect to introduce human factors into the initial design discussions for the "symbio-bot." Such techniques exist in the form of serious gaming [14, 15]. In that sense, serious gaming can be used to introduce the concept of the "symbio-bot" to an audience of people who should not necessarily have a scientific or technical background. This audience is intended to provide information concerning the interface between the "symbio-bot" and the patient. The aim of such serious gaming is to assess the "symbio-bot" at the level of the paradigm concerning the acceptance and trust by patients that the "symbio-bot" will function successfully in their body. Clearly, the serious gaming needs to have a good proposal for the "symbio-bot" and a good virtual representation of the function of the "symbio-bot" and how it would interact with the body. This means that the "symbio-bot" already needs to have a good level of technical specification of the interface with tissues of the body but provided at the level that can be implemented by the human factors' experts in an appropriate form for serious gaming. This could be in the form of a computer simulation, or a 3D-printed prototype model, or a well-illustrated set of images.

The importance for the development of the "symbio-bot" is to have information about its acceptance by the patient, or indeed an audience of non-technical end-users. Clearly this type of information could be a "yes/no" for a "Go/Nogo" for the initial concept of the "symbio-bot." Even if there is scientific merit at this early stage of development, unless the design of the "symbio-bot" achieves acceptance by patients then the further complex technical development will not provide a successful final outcome for the "symbio-bot."

At this point in the discussion, we can consider the ways in which serious gaming is used generally in the development of medical interventions and technical disciplines. Then we can draw some direction as to how serious gaming might be used in the case of the "symbio-bot." The utilization of serious gaming in the biotechnology field was first published by Riedel-Kruse et al. [16], who proposed "biotic games" that operate on biological processes. Games are a significant and defining part of human culture, and their utility beyond pure entertainment has been demonstrated with so-called "serious games." Biotechnology, despite its recent advancements, had no impact on gaming prior to 2011. Utilizing a variety of biological processes, Riedel-Kruse et al. [16] designed and tested a collection of games: "Enlightenment," "Ciliaball," "PAC-mecium," "Microbash," "Biotic Pinball," "POND PONG," "PolymerRace," and "The Prisoner's Smellemma." They found that biotic games exhibit unique features compared to existing game modalities, such as utilizing biological noise, providing a real-life experience rather than virtual reality, and integrating the chemical senses into play. Analogous to video games, biotic games could have significant conceptual and cost-reducing effects on biotechnology and eventually healthcare; enable volunteers to participate in crowdsourcing to support medical research; and educate society at large to support personal medical decisions and the public discourse on bio-related issues [16].

The utilization of serious gaming in digital health and in medical treatments dates from 2012, when the first publications were aimed at the educational needs of doctors in preparing for medical practice [17] or for complex interventional techniques [18]. As an example, when applied to "digital health," in 2016 the advantages of innovations in digital health were envisioned as a world where clinicians would learn hemodynamics with simulators and serious games, would monitor patients with wearable or implantable sensors in the hospital and after discharge, would use medical devices able to communicate and integrate the historical, clinical, physiologic, and biological information necessary to predict adverse events, propose the most rational therapy, and ensure it is delivered properly. Considerable intellectual and financial investments are currently made to ensure some of these new ideas and products soon become a reality [19].

In the field of implantable medical devices, Gnadlinger et al. [20] reported a recent application of serious gaming for patients who receive cochlear implants. Cochlear implants are implanted hearing devices; instead of amplifying sounds like common hearing aids, this technology delivers preprocessed sound information directly to the hearing (i.e., auditory) nerves. After surgery and the first cochlear implant activation, patients must practice interpreting the new auditory sensations, especially for language comprehension. This rehabilitation process is accompanied by hearing therapy through face-to-face training with a therapist, self-directed training, and computer-based auditory training. It was demonstrated as highly feasible to combine an intelligent tutoring system with a serious game in cochlear implant rehabilitation therapies. This could lead to further advances in cochlear implant care and aural rehabilitation in general [20]. Çetinkaya et al. [21] applied the approach of serious gaming to the difficult situation of a cochlear implant used to enhance the hearing of young children between the ages of 3 and 5 years. They developed a serious game called "World of Sounds," which included four components: *(i)* the crucible sound for auditory awareness, *(ii)* mole hunting for auditory discrimination, *(iii)* find the sound for auditory recognition, and *(iv)* choo-choo for auditory comprehension. A study that included the parents of a group of children with normal hearing compared to a group of children with cochlear implants showed some usefulness of this serious game approach for the auditory rehabilitation of children with cochlear implants [21]. Furthermore, an earlier randomized controlled trials of Reynard et al. [22] assessed serious gaming as a rehabilitation tool to improve speech-in-noise intelligibility in adult users of cochlear implants. The results suggested that serious game-based auditory training may improve speech-in-noise intelligibility in adult users of cochlear implants. Serious gaming may be considered as a complementary training approach for improving outcomes in adults who use cochlear implants [22].

From those experiences with serious gaming, particularly as exemplified by its usefulness for patients to better adapt to cochlear implants, we can consider the advantages that serious gaming may have when applied to "symbio-bots." To underline those advantages, we can return to the opening statements made in this chapter about redefining the discipline if biological engineering. This refined discipline of biological engineering includes the aspect of human factors concurrently with technical factors, particularly when we consider the therapeutic and diagnostic needs for nanomedicine. Such needs rely upon medical devices, either implanted or in contact with the body of a patient. The enhancements to such medical devices will require them to be "smart," in the sense that there will need to be an improved understanding and construction of such devices to provide for long-term two-way (duplex) communication of biological molecules and signals to and from the device to the body. Concurrently, the patient needs to be accepting of such innovative "smart" devices that will have the capability to function autonomously. That is, the patient needs to understand and provide permission for the "smart" device to function autonomously. Here we return to the roadmap presented in Chap. 2 (Fig. 2.14) and the six questions that are at the heart of the *"medical device pull"* to utilize skills of materials and engineering disciplines to design the "symbio-bot" that is required to solve the clinical needs.

We can consider using the use of serious gaming as a form of understanding and testing the "man-machine interface," but adapted at the level of the implanted "symbio-bot" medical devices that have the technical capabilities to function autonomously inside the body. The reasons for drawing such a parallel are based on the idea that the way a person uses a technical device is an important influence on the design of that technical device. These concepts gave birth to the formal discipline of ergonomics, which is an important aspect of good engineering design.

Lessons from Ergonomics in Framing the Paradigm Shift for Biological Engineering

A decade ago, the predicted growth of ergonomics as a scientific discipline was considered to depend on the way its practitioners were able to transfer ergonomic findings that arose from the research field into ergonomic application in the companies [23]. This highly practical view of ergonomics most likely was deeply rooted in its origins as an engineering discipline for improving man-machine interactions, typically at an industrial level. An approach of human factors/ergonomics (HFE) has several methodological challenges when applied to the design of complex work systems [24]. Norros [24] utilized experience from the VTT Technical Research Centre of Finland, which is an interdisciplinary national research institute positioned between Finnish academia and industry, to show how HFE has been applied mainly to improving the operations of plants and organizations. In 2014, Norros described the need for involving HFE in the design of tools and technologies, or in the planning of future operations. In that context, it was important to consider the unity between human technology and environment. This led to the concept of *technology-in-use* as the perspective to bring technological and human elements together as inseparable components. The three practical ways of considering this *technology-in-use* were first to understand that technology is a tool that people take advantage of in their various activities, second to understand that technology is a medium that mediates messages for people that are incorporated in the technology itself, and third that technology is something that needs to be taken care of since people use technology to shape their environment and thus the environment becomes a resource of people's activity [24]. Those *technology-in-use* concepts were clearly motivated by the need to understand the use of technology in the work activities of people and to improve the design of technology by incorporating HFE into engineering design.

Beyond safety, further objectives, such as the usability of tools and services and experience concerning their use, have also been identified by Norros as significant motivators for exploitation of HFE [24]. Reaching the usability-driven objectives is considered to have an influence on the performance of the systems. This potential is demonstrated by the interest that the concept of "User Experience" (UX) has raised among some leading companies of the Finnish and international metal industry, resulting in the launch of a large public/private partnership type of research program [25]. The emergence of the UX construct characterizes the so called third wave usability research, and it has, in particular, been connected to new business possibilities that the human-centered design could provide [26]. Usability and UX research originate in the Human-Computer Interaction tradition that has the advantage of having an intimate connection to design activity [27].

Although those concepts of HFE and UX were framed to improve the design of technology by people, we can adopt some of those concepts in considering the design of "symbio-bots." In particular, the usage of modelling methods, simulations or virtual techniques to experiment together with the users to anticipate functioning of "symbio-bots," to make implicit knowledge explicit, and to create dialogue among the different disciplines that are needed to produce "symbio-bots."

Indeed, a recent formalisation of Ergonomics-in-Design is targeted to evaluate and design the global quality of systems which people enter into a relationship with during their work activities and daily life [28]. This is an interesting *point of departure* when considering "symbio-bots." We can consider ergonomics in design as a *point of departure* since its definition positions the systems as external to people but clearly important for significant interaction with people. The "symbio-bot" is a further and significant extension of this Ergonomics-in-Design concept, since the "symbio-bot" functions autonomously of a person, but after the person has accepted the usage of a "symbio-bot" that is integrated with their body following a medical procedure. That is, following the initial integration of the "symbio-bot," the person has no daily choice of whether or not to use the "symbio-bot" since it will be deployed either by implantation or by long-term connection with the skin. In the sense of a long-term connection with the skin, the person needs to understand and accept that the "symbio-bot" has been properly approved under some form of regulatory affairs process. To adopt the jargon of Ergonomics-in-Design, the "symbio-bot" needs to be the outcome of Human-Centered Design, which describes a design approach that is aimed at the quality of the interaction between people and systems they come into contact with, based on collecting and processing essential data so as to understand people's needs through structured and verifiable methods for study and evaluation. The need to integrate human factors into good design of medical devices has been known by scientists, engineers, and clinicians as part of common sense for good engineering and healthcare practice. Nonetheless, it is essential to have a formal framework, such as was described by Vincent et al. [29] to better understand the constraints under which medical device design and development take place [29].

Central to this approach of Human-Centered Design is the need to identify appropriate users for the UX, or indeed validated personas for medical equipment design [30]. We can define a persona as a composite archetypical profile of the main users of a technology, without specifying any particular individual person [31]. Indeed, personas provide ways of thinking about how users behave, think, what they want to accomplish, and why. They are composite archetypes that stop developers from falling into the trap of designing for themselves [29]. The need to define personas is also a central aspect of the need for explainability in the use of artificial intelligence (AI) in healthcare. Despite the potential benefits of AI and deep learning in assisting doctors to make diagnoses, there is a lack of transparency in understanding how the AI algorithms function to make decisions. This diminishes the level of trust in AI when applied to healthcare. To remove this transparency in understanding how AI reaches decisions, the concept of explainability has been formally defined, with Dey et al. [32] discussing how user-driven explainability can be made more useful for different healthcare stakeholders through the definition of three key personas: being data scientists, clinical researchers, and clinicians [32]. The biggest challenge to design a model for explainability in AI is to understand how a human might expect an explanation to be. For example, a medical professional society might want a description of the behavior of the entire model, a global explanation, in order to inform best practices for their membership. In contrast, an individual clinician is required to provide a local explanation to a patient who wants to know why the AI model has predicted that he/she has/may have cancer.

An analogous situation for explainability would arise for the implantation of a "symbio-bot." The "symbio-bot" has autonomous control of its function when implanted, which increases the level of difficulty for explainability. The implantation of the "symbio-bot" raises the level of difficulty in explainability since the patient would be required to make an informed decision about implantation based on an explanation from a clinician who may not be completely knowledgeable about the technical capabilities of an autonomous "symbio-bot." It is important to define the appropriate personas for modeling the explainability. The utilization of serious gaming would provide an ideal environment to ensure all personas have the appropriate explanations of the autonomous function of a "symbio-bot." To best prepare for such utilization of serious gaming, it is important to consider the background of integrating human factors and the user experience (UX) into the serious gaming. The rapid increase in accessibility to social media provides an intriguing implementation of the UX. Winterstein et al. [33] demonstrated the remarkable potential of combining generative AI with strategically targeted advertisement using social media for digital health communication. A persona, named "Wanda," was created using generative AI (Midjourney, www.midjourney.com) and utilized as a virtual influencer to transmit cancer prevention messages to the public on Instagram. This also highlighted the potential of virtual AI-driven influencers as scalable tools for digital health communication [33].

The concept highlighted by Winterstein et al. [33] can be utilized to create virtual generative-AI personas, which would be incorporated into a UX for the explainability of the "symbio-bot" function when implanted. Such a generative AI-driven approach provides enormous potential scalability and an easy means of outreach to several target groups of users for communicating the function and benefits of a "symbio-bot." Such utilization of generative AI to create virtual personas would enhance the codesign for a "symbio-bot." The principles of codesign are to highlight the needs and preferences of the consumer or end-user, which is the patient in the case of a "symbio-bot." Successful codesign of a "symbio-bot" would need the three elements of *(i)* explanation of how it functions, *(ii)* understanding that the "symbio-bot" functions under autonomous control inside the body, and *(iii)* acceptance by the patient of the autonomous function of the "symbio-bot." Creating virtual generative AI personas for wide distribution on social media would provide a patient the means to both participate in the codesign and to be informed by explanations of the functions of the "symbio-bot." Such utilization of virtual personas and social media would encapsulate the principles of the UX codesign approach as highlighted by Nesbitt et al. [34], *(i)* the design is based on a sound understanding of the users, tasks, and environments under review; *(ii)* there is user participation throughout the whole process; *(iii)* design is driven and refined by user-centered evaluation; *(iv)* the process is iterative; *(v)* the design addresses the whole user experience; and *(vi)* the design team is multidisciplinary and poly-perspectivity [34].

A related example, which does not utilize virtual generative AI persons but does encompass the principles of UX codesign, is a study reported by Yu et al. [35] into the use of a new type of virtual reality (VR) headset. The study aimed to develop questionnaires for evaluating VR glasses, as well as to investigate the relationship between various UX variables. Using lab-based usability tests, this study analyzed participants' self-reports and performance based on testing eight VR glasses and seven mobile applications. A nine-item questionnaire and a ten-item questionnaire were successfully developed to measure VR glasses systems' UX quality in terms of hardware and application, respectively. Within a proposed UX evaluation framework, the perceived UX quality relative to VR glasses hardware emerged as a core predictor in predicting interactive operation performance, whereas the application UX perception was a significant predictor of motion sickness [35].

Concluding Remarks

We can now return to the context of how these concepts of human factors can be included in the technical development of a "symbio-bot." Here we need to introduce the needs for the Societal Readiness Level (SRL) to be evaluated concurrently with the Technology Readiness Level (TRL). The significant opportunity in framing this new health paradigm for "symbio-bots" is to challenge all the scientific disciplines of humanities and social sciences (SHS). This is particularly important in constructing the philosophical and ethical framework for developing expertise in UX design, including codesign with patients, and in corporal/emotional communication and design. The explicit response for "symbio-bot" is the process of understanding and appropriating this new paradigm for patients, to then demonstrate the relevance of multidisciplinarity at theoretical and methodological complementarity. A challenge is to demonstrate whether technologies like "symbio-bots" transform the habits and expectations of patients, which have been traditionally built around the natural limits of the human being and body that are intuitively obvious to people. Such limits are also shifted so that a new paradigm of symbiotic medical devices, such as "symbio-bots," takes on a meaning with regard to a realization of the model of the vision of medicine as predictive, preventive, personalized and participatory ("P4 medicine").

References

1. http://ec.europa.eu/growth/sectors/medical-devices_en
2. Deloitte. (2022). Global Health Care Outlook: Are we finally seeing the long-promised transformation? https://www2.deloitte.com/global/en/pages/life-sciences-and-healthcare/articles/global-health-care-sector-outlook.html
3. https://www.snitem.fr/

4. Schuijff, M., & Dijkstra, A. M. (2020). Practices of responsible research and innovation: A review. *Science and Engineering Ethics, 26*, 533–574.
5. European Commission, Directorate-General for Research and Innovation. (2011). In R. Schomberg (Ed.), *Towards responsible research and innovation in the information and communication technologies and security technologies fields*. Publications Office. https://data.europa.eu/doi/10.2777/58723
6. Alcaraz, J., Cinquin, P., & Martin, D. K. (2018). Tackling the concept of symbiotic implantable medical devices with nanobiotechnologies. *Biotechnology Journal, 13*, 1800102.
7. El Ichi-Ribault, S., Alcaraz, J. P., Boucher, F., Boutaud, B., Dalmolin, R., Boutonnat, J., Cinquin, P., Zebda, A., & Martin, D. K. (2018). Remote wireless control of an enzymatic biofuel cell implanted in a rabbit for 2 months. *Electrochimica Acta, 269*, 360–366.
8. Cinquin, P., & Martin, D. K. (2017). *Biomimetic artificial membrane device*. US 9,577,280.
9. Thélu, J., Cinquin, P., Martin, D. K., Soranzo, T., Alcaraz, J. P., & Kweter, J. M. (2021). *Device for collecting intestinal fluid*. EP 3,700,431 B1.
10. Cinquin, P., Schneider, D., Maurin, M., Martin, D. K., El Ichi, S., Zebda, A., Alcaraz, J. P., Reche, F., Tuvignon, P., Thelu, J., Le Gouellec, A., & Toussaint, B. (2021). *Implantable intestinal reactor*. US 10,960,191.
11. Martin, D. K., & Picollet-D'hahan. (2022). *Method for diagnosing genitourinary cancers*. US 11,268,961.
12. Martin, D. K., & Picollet-D'hahan. (2024). *Cell co-culture chip and process for the production thereof*. US 11,857,958.
13. Van der Sanden, B., Gredy, L., Wion, D., & Olivier, S. (2021). 3D two-photon polymerization of smart cell gelatin – Collagen matrixes with incorporated ruthenium complexes for the monitoring of local oxygen tensions. *Acta Biomaterialia, 130*, 172–182.
14. Gillois, P., Di Marco, L., Nicolàs, J. M., Moreau-Gaudry, A., Ego, A., David-Tchouda, S., Martin, D. K., & Bosson, J. L. (2020). Biostatistics disruptive acculturation through serious gaming: A new hope. *Studies in Health Technology and Informatics, 270*, 1215–1216.
15. Di Marco, L., El Kechaï, H., Martin, D. K., & Gillois, P. (2019). Blended learning for French health students: Does acceptance of a learning management system influence students' self-efficacy? *Studies in Health Technology and Informatics, 264*, 1169–1173.
16. Riedel-Kruse, I. H., Chung, A. M., Dura, B., Hamilton, A. L., & Lee, B. C. (2011). Design, engineering and utility of biotic games. *Lab on a Chip, 11*(1), 14–22.
17. Hannig, A., Kuth, N., Özman, M., Jonas, S., & Spreckelsen, C. (2012). eMedOffice: A web-based collaborative serious game for teaching optimal design of a medical practice. *BMC Medical Education, 12*, 104.
18. Chan, W. Y., Qin, J., Chui, Y. P., & Heng, P. A. (2012). A serious game for learning ultrasound-guided needle placement skills. *IEEE Transactions on Information Technology in Biomedicine, 16*, 1032–1042.
19. Michard, F. (2016). Hemodynamic monitoring in the era of digital health. *Annals of Intensive Care, 6*, 15.
20. Gnadlinger, F., Werminghaus, M., Selmanagić, A., Filla, T., Richter, J. G., Kriglstein, S., & Klenzner, T. (2024). Incorporating an intelligent tutoring system into a game-based auditory rehabilitation training for adult cochlear implant recipients: Algorithm development and validation. *JMIR Serious Games, 12*, e55231.
21. Çetinkaya, M. M., Konukseven, O., & İralı, A. E. (2024). World of sounds (Seslerin Dünyası): A mobile auditory training game for children with cochlear implants. *International Journal of Pediatric Otorhinolaryngology, 179*, 111908.
22. Reynard, P., AttinaV, I. S., Hermann, R., Barilly, C., Veuillet, E., Joly, C. A., & Thai-Van, H. (2022). Effect of serious gaming on speech-in-noise intelligibility in adult cochlear implantees: A randomized controlled study. *Journal of Clinical Medicine, 11*, 2880.
23. Vidal, M. C., Guizze, C. L. C., Bonfatti, R. J., & Silva e Santos, M. (2012). Ergonomic sustainability based on the ergonomic maturity level measurement. *Work, 41*, 2721–2729.

24. Norros, L. (2014). Developing human factors/ergonomics as a design discipline. *Applied Ergonomics, 42*, 61–71.
25. FINAL REPORT: UXUS. *User experience and usability in complex systems.* November 17, 2015. https://www.dimecc.com/final-report-uxus-user-experience-and-usability-in-complex-systems/
26. http://drops.dagstuhl.de/opus/volltexte/2011/2949/pdf/10373_AbstractsCollection.2949.pdf
27. Savioja, P., & Norros, L. (2013). Systems usability framework for evaluating tools in sfety-critical work. *Cognition Technology and Work, 15*, 255–275.
28. Tosi, F. (2020). Ergonomics in design and design in ergonomics. *Work, 66*, 909–911.
29. Vincent, C. J., Li, Y., & Blandford, A. (2014). Integration of human factors and ergonomics during medical device design and development: It's all about communication. *Applied Ergonomics, 45*, 413–419.
30. Vincent, C. J., & Blandford, A. (2014). The challenges of delivering validated personas for medical equipment design. *Applied Ergonomics, 45*, 1097–1105.
31. Goode, A. D., Frith, M., Hyne, S. A., Burzic, J., & Healy, G. N. (2022). Applying a user centred design approach to optimise a workplace initiative for wide-scale implementation. *Environmental Research and Public Health, 19*, 8096.
32. Dey, S., Chakraborty, P., Kwon, B. C., Dhurandhar, A., Ghalwash, M., Suarez Saiz, F. J., Ng, K., Sow, D., Varshney, K. R., & Meyer, P. (2022). Human-centered explainability for life sciences, healthcare, and medical informatics. *Patterns, 3*, 100493.
33. Winterstein, J. T., Abels, J., Kuehn, A., Carl, N., Wies, C., & Brinker, T. J. (2025). AI-generated cancer prevention influencers can target risk groups on social media at low cost. *European Journal of Cancer, 217*, 115251.
34. Nesbitt, K., Beleigoli, A., Du, H., Tirimacco, R., & Clark, R. A. (2022). User Experience (UX) Design as a co-design methodology: Lessons learned during the development of aweb-based portal for cardiac rehabilitation. *European Journal of Cardiovascular Nursing, 21*, 178–183.
35. Yu, M., Zhou, R., Wang, H., & Zhao, W. (2019). An evaluation for VR glasses system user experience: The influence factors of interactive operation and motion sickness. *Applied Ergonomics, 74*, 206–213.

Chapter 4
Where Does Biomimetic Membrane Technology Fit into Biological Engineering?

The phospholipid membrane is one of biology's most pervasive structures and represents an ideal scaffold for a host of nanotechnology applications. Whether as a vehicle for the biological engineering of biomimetic technologies or for designing therapies to interface with the cell, a phospholipid membrane can provide the necessary molecular-level control of membrane-anchored proteins, glycopeptides, and glycolipids applications [1]. An elegant published example of a lipid membrane structure that illustrates this concept is the incorporation of two proteins that have a combined function to produce ATP [2]. This structure includes the proteins bacteriorhodopsin and the F_0F_1-ATPase complex. The structure is illustrated in Fig. 4.1.

Wendell et al. [2] described the function of the natural structure illustrated in Fig. 4.1 as commencing with solar radiation stimulating bacteriorhodopsin to translocate protons across the membrane, which produces a proton gradient. Then, the F_0F_1-ATPase complex utilizes the gradient to pump the protons into the cell, stimulating the synthesis of ATP. Formation of purple membrane and BR are light-dependent processes. The absorption spectra of BR range from 400 to 700 nm, with a peak absorbance at approximately 580 nm (yellow-green light). ATPase is a ubiquitous enzyme, with the two major portions of (i) F_0, the hydrophobic, membrane-bound portion that is responsible for proton translocation and (ii) F_1, the hydrophilic portion that is responsible for ATP synthesis and hydrolysis. As protons flow through the F_0, the γ subunit rotates clockwise and ATP is synthesized. a, b, and c subunits of the F_0-ATPase form the channel that allows protons to flow through the membrane. Artificial systems consisting of bacteriorhodopsin and F_0F_1-ATPase in liposomes have been used to demonstrate the light-driven production of ATP [3, 4]. In these systems, bacteriorhodopsin pumps protons from the surrounding media into the liposomes. The gradient is then utilized by F_0F_1-ATPase to synthesize ATP external to the liposome and transport protons back across the liposome membrane.

D. Martin, I. Vilgrain, *Biomimetic Membrane Technology*,
https://doi.org/10.1007/978-3-031-90594-0_4

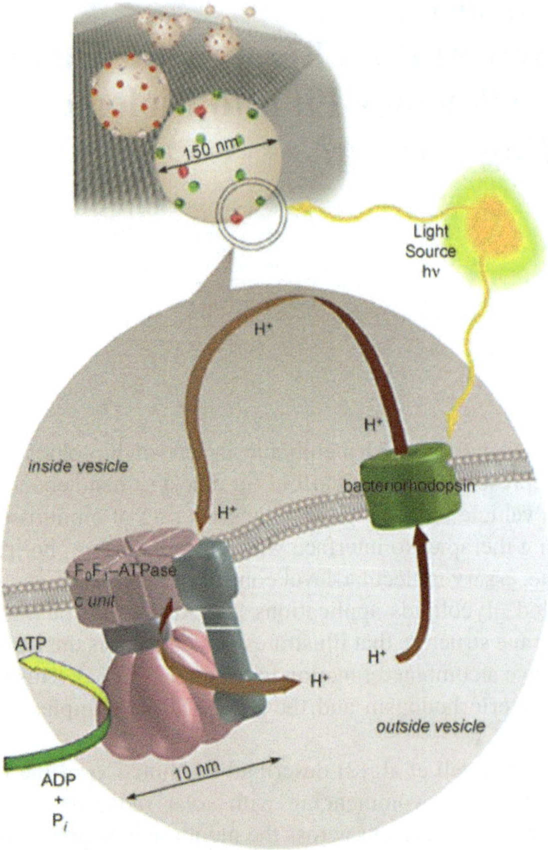

Fig. 4.1 Coupled ATP production with F0F1-ATPase and bacteriorhosopsin. (Reproduced with permission from [2])

The example illustrated in Fig. 4.1 illustrates a metaphor for the use of lipid bilayer membranes as a foundation for smart nanostructured systems that are developed using biological engineering. The assembly of proteins in lipid bilayers provides components that give function to such smart nanostructured systems. A useful context for such elegant lipid membrane-based nanostructures is to enhance the function of a "symbio-bot" to improve the diagnosis or therapy, particularly for in vivo applications. We explore the use of the lipid bilayer membrane to provide a more appropriate interface between a functional implanted device and the organs of the body. We also explore here the technical role for nanobiotechnologies in diagnostics and therapeutics, then return to the question of the role of the patient, when armed with appropriate information, as being part of the diagnostic and therapeutic team. Explaining this context to the patient helps to understand the concept of how biological engineering utilizes natural systems to enhance a device that is biocompatible and functions as a diagnostic or therapeutic system.

A Role for Nanotechnologies to Enhance the Function of Diagnostic and Therapeutic Devices

The word nanotechnologies immediately conjure up visions of nanoparticles swarming rampantly around and into our bodies. To best prepare a patient to understand the role that nanotechnologies can play in diagnosis and therapy, it is important to identify the broader applications of nanotechnologies and biological engineering that enhance the functions of diagnostic and therapeutic devices. The patient will be better prepared to accept and participate in the co-design of innovative "symbio-bots" if there is meaningful pedagogy that nanotechnologies provide far more than only nanoparticles.

Nonetheless, as we examined in Chap. 1, the predominant themes in the existing nanomedicine literature include nanoparticles and drug delivery. In 2012, the literature reiterated the difficulty in constructing drug-loaded NPs with precisely defined therapeutic effects, in terms of release time and spatial targeting, but offered a perspective to improve the nanoparticle-based drug delivery [5]. Gong and Winnik [5] reported that a particularly successful design strategy consisted of masking the nanoparticle to the biological environment by covering it with an outer surface that mimicked the composition and functionality of the cell's external membrane. To achieve such a biomimetic surface engineering of nanoparticles, the key challenges to be overcome were identified as (i) avoiding toxicity to healthy cells and improving the targeting of the nanoparticles by incorporating several targeting ligand molecules in the surface engineering, (ii) optimizing the density of targeting ligands in the surface engineering, and (iii) improving and controlling the structure of the engineered surface. A pathway for the surface engineering was proposed to use amphiphilic copolymers that comprised phosphorylcholine and other groups that could be functionalized to enhance the biocompatibility of the nanoparticles [5].

That concept of biofabricating an interface between a device and biological tissue can be traced back to the Stone Age and that prehistoric people were aware that the body itself had capabilities for self-restoration, without a real understanding about biocompatibility issues [6]. A more formal engineering perspective to biofabricating the (bio-device) interface between a device and biological tissue had been published in 2010 by Liu et al., even if the approach was to add biological components to an electronic device after it had been manufactured [7]. In that publication, the vision of biofabrication aimed to generate a bio-device interface that was not permanent so the chip/device can be washed, re-functionalized, and reused. Liu et al. [7] envisioned devices that can perform functions (e.g., analysis) while fully accessing the power of biology (for molecular recognition) and electronics (for data collection, analysis, and transmission), while device reusability would allow these functions to be performed economically.

Even if those publications provide only a limited overview into the extensive literature on nanoparticle-based drug-delivery and therapeutics, and for bio-device interface engineering, those publications do highlight the added value of considering the interface as an essential aspect of any nanotechnology-based diagnostic or therapeutic system. Now, more than 20 years after the publication of Liu et al. [7], the interface continues to remain the subject of research efforts to control this important feature.

Adding to that concept of surface engineering to enhance biocompatibility and function, Goes and Fuhrmann [8] wrote that biogenic approaches offer the potential to counterbalance the drawbacks of artificial carriers, as they may naturally overcome barriers associated with anti-infective therapy. They defined a biogenic approach as one that is produced or originates from a living organism, such as a cell, extracellular vesicle, or an enzyme. Biogenic avenues provide exciting opportunities to develop novel future concepts, by combining natural and artificial systems. Nonetheless, they identified a bottleneck as being the isolation and preparation of such nature-derived approaches in a reproducible and upscalable manner [8].

Moving forward to 2022, Guimarães et al. published an opinion overview of engineering living bioassemblies for biomedical and functional medical applications. Of relevance to our discussion here, the authors provided an outlook that identified several remaining challenges but without identifying a clear solution to the problem of the interface between the device and biological tissues [9]. Those challenges are summarized in Box 4.1.

Box 4.1: Challenges Identified for the Engineering of Living Bioassemblies for Biomedical and Functional Material Applications, as Identified in the Outlook for the Field Published by Guimarães et al. [9]

- The diversity of fabrication technologies will require the development of universal standardized characterization and reporting protocols to compare and evaluate the performance and viability of bioassembled constructs.
- The use of machine-learning approaches could potentially predict and suggest the desired combination of building blocks (i.e., bacteria, cells, biological matrices) to mitigate the endless possibilities of combinations and designs in order to achieve particular functionalities or material properties.
- The fabrication of living tissues using human cells and engineered matrices could also accelerate the translation of innovative biotechnology advances from the lab bench to the clinical settings, complementing and potentially significantly reducing the need for animal testing.
- The potential for a very complex regulatory framework and extensive market regulation could be a strategically significant challenge for the translational potential of biomedical application-related bioassemblies.
- Identification of commercially viable applications outside of biomedical applications could help accelerate the translation of bioassemblies into everyday technologies.
- Limitations of scale-up for specialized biological building blocks (cells, bacteria, and biological matrices) could be mitigated by using compatible, simple, and programmable culture conditions that would potentially enable rapid adoption.
- The major bioethical challenge that arises from expanding tissue engineering to create engineered organisms, since the use of bioassemblies could have a profound impact on our everyday lives and at the clinic, ranging from extended life expectancy to healthier sustainable foods and environmentally sustainable materials.

The identification of such challenges helps to guide the development of diagnostic and therapeutic devices from a technical perspective but can make the development difficult to understand for the patient who needs to be an active participant in the team. To help provide the patient with appropriate information, the question of the interface between the device and biological tissues could be approached from a general biological perspective, rather than initially from the technical needs of the medical device. For such a general perspective, the implantation of device or material in a body was conceptualized by Alcaraz et al. [6] using the simple image of communication between two different countries separated by a frontier [6]. Alcaraz et al. [6] described the scenarios where those two countries need to maintain good relationships in order to avoid a war. In that scenario, we could consider a biodegradable implanted material as being a country that provides a useful one-way communication to the second country. As this country soon disappears (i.e. the biodegradable implanted material), then the frontier no longer is a barrier, the body (the other second country in the scenario) cures, and the war effectivly disappears. Taking a regenerative medicine approach to implant several stem cells can lead to cell proliferation if the environment inside the body is not well controlled, in a sense because there is no well-defined frontier to contain the stem cells even though there is communication between individual stem cells and the body. Although there is not a frontier, the implanted stem cells will win the war if the resulting tumor is not removed. A more complex implanted medical device usually stimulates the building of a wall at its frontier, as a consequence of the inflammatory reactions of the body to encapsulate the implanted device. The inflammatory responses will lead to "cold war" between the implant and the body, reducing the communication of substances, and after varying periods of time, the implant will cease to be of benefit to the patient. The inflammatory reactions can be reduced by drugs such as cyclosporins, which may reduce substantially the wall's barrier functions but also induce adverse effects that reduce the intended two-way communication to a whisper for an implanted device that is intended to be symbiotic.

In those conceptual examples described by Alcaraz et al. [6], implant signaling (one-way or simplex) resulting from the presence of the exogenous material itself, an electric signal or a drug delivery leads to two consequences: the disappearance of the biomaterial (biodegradation) or encapsulation of the implanted material. The two-way communication (or duplex) following the implantation of a symbiotic device provides for dynamic interactions between the frontier of the symbiotic device and the physiological environment of the body. Good duplex communication minimizes encapsulation, which could only be obtained without immunosuppressive drugs by autologous grafts.

Clearly, this is an observation and not a requirement. For example, a pregnant woman does not take immunosuppressive drugs although foreign cells are present for 9 months in her body and exchange molecules (i.e., duplex communication). The absence of rejection of the fetus remains a controversial topic of pregnancy. Indeed, the mother rejects any tissue transplant from her child after birth, whereas she accepted this natural "allograft" for several months. Nonetheless, even miscarriage happens and the etiology of recurrent miscarriage is unknown in about 50% of cases. Despite the hypothetical role of natural killer cells in decidualization, the role played by alloimmune mechanisms remains unclear. Several hypotheses have been put forward to explain recurrent miscarriage, such as HLA-C-specific antibodies, the memory CD8-T cell population, or even the immunoregulatory effect of T-cell immunoglobulin and mucin domain-containing protein 3 (TIM3). It has been proposed recently that inflammatory processes implicated in physiological human reproduction are triggered mainly through sterile inflammation pathways, via alarmins signaling.

In 2021, a publication by Govindasamy et al. described a 3D biomimetic environment that was used to investigate the cellular mechanisms of underlying embryo-maternal interactions during implantation [10]. This publication provides a device for examining a complex biological interface, which also provides a biological perspective to the issue of explaining the bio-device interface. Although every in vitro model has the limitation of reducing highly complex in vivo conditions to an approximation, the 3D model of Govindasamy et al. [10] provided an environment that revealed dynamics of the embryo's first contacts with maternal vasculature. This type of 3D model of an important interface could provide the type of information in a form that is accessible to the patient to improve their understanding of the bio-device interface.

Indeed, in this book, we emphasize the significance of such an interface as providing the frontier to be conquered for improving the function of nanostructured devices. In Chap. 1, we introduced the concepts of biological engineering, and particularly biometric membrane technology, as emerging disciplines that introduce a paradigm shift in the way that nanobiotechnologies are used in nanomedicine. The paradigm shift of this new field of biological engineering is that a nanostructured biomimetic membrane is used in medical devices to provide an interface that is self-assembled in similar ways to the cells of the body. This provides the medical device with an interface that will be capable of communicating biological molecules with the body. The medical device will have the capability to function autonomously to assist the functions of cells and tissues in the body. This paradigm shift implicates that the framing of the problem is the driver for the multidisciplinary research needed to provide a solution, utilizing interdisciplinary skills that are embodied in the emerging field of biological engineering. Such a concept leads us to consider the benefits that nanotechnologies can bring to the field of nanomedicine.

Benefits for Nanomedicine from Nanobiotechnologies Based on Biomimetic Membrane Technology

In digging deeper into the rationale for using a lipid bilayer membrane to provide a more appropriate interface between a functional implanted device and the organs of the body, in this section, we examine several aspects of biomimetic membrane technology that can provide benefits for nanomedicine. These aspects are drawn from the literature to illustrate the breadth of research that is stimulated by the advantages of nanobiotechnologies based on biomimetic membrane technology. The discussion tries to focus on lipid bilayer membranes, but it is worth reiterating that for most people, the field of nanotechnologies continues to invoke the notion of nanoparticles. Even if nanomedicine is benefitted by nanotechnologies and bioinspiration, it seems that these benefits are directed toward enhancing the performance of nanoparticles. For example, taking the cell membrane as the inspiration for nanoparticle design brings the strategies of incorporating components of the cell membrane onto nanoparticles to enhance targeted delivery applications [11].

Benefits for nanomedicine from nanobiotechnologies also include the utilization of nanostructured biomimetic membranes that are not based on lipids. This utilization touches on tissue engineering and artificial cell applications. One of the most versatile and simplest biomimetic membranes adsorbed to devices utilizes the concept of semipermeable films assembled from polyelectrolytes, which was introduced in 1960 by Mindick and Patwelt [12]. The development of the layer-by-layer (LbL) technique of assembling polyelectrolyte films by Decher and Hong in 1991 stimulated investigation of the use of the films for many applications [13, 14]. The LbL manufacturing process involves the alternate adsorption of polyelectrolytes of opposite charge onto surfaces including plastic, glass, latex, colloidal particles, bacterial or mammalian cells. Polyelectrolyte multilayer films have provided a permeable substrate to support lipid bilayers but initially were adsorbed onto a solid supporting structure and were not freestanding [15, 16]. We have described a novel way of forming a free-standing planar polyelectrolyte multilayer film using a sacrificial substrate of low melting point agarose to produce an interpenetrating hydrogel (IPH) [17]. The production of free-standing LbL multilayer films has definite advantages for diffusion membranes [18]. The simplicity and versatility of the LbL technique, especially when modified to produce hollow microcapsules [19], allows for many potential biomedical applications such as micro-encapsulation, controlled drug release, biosensors, as well as tissue engineering. We have recently published a biological application with our approach utilizing the LbL technique to create a free-standing hollow microcapsule scaffold to support an ion transporting lipid bilayer for application in an artificial cell [20].

Fundamental Nanostructure of a Tethered Lipid Bilayer Membrane Without Incorporated Proteins

We turn to biomimetic membranes assembled from lipids. The advantage of these membranes is that proteins can be incorporated to provide functionality to the membranes. Such functionality can include ion transport, biosensing, biocompatibility, and other mimics of biological function. These types of functional biomimetic membranes provide excellent building blocks for the biological engineering of symbio-bots that can be designed for diagnostic or therapeutic devices for nanomedicine. The advantage of biomimetic membranes as building blocks is their apparently limitless potential for utilization in nanomedicine. However, rational engineering requires detailed knowledge of the structure and properties of the building blocks to be utilized. The structure and properties of proteins and lipids are at the nanoscale and so present challenges to first obtain detailed information that can then be used in biological engineering at the nanoscale.

The fabrication of model lipid membranes possessing some degree of predictability was first reported in the pioneering work of K.R. Blodgett and I. Langmuir who formed thin monomolecular films supported on solid or liquid surfaces [21–23]. Prior to that pioneering work, the thickness of suspended bimolecular liquid films of soap in air had been investigated and measured [24, 25]. The extension of that work to produce bimolecular liquid films in air led to an understanding of the formation of lipid films in water [26]. Based on those early reports, a major milestone in suspended lipid bilayer research was achieved by the reconstitution of an excitable cell membrane structure in a saline solution, which was in-effect the first report of the production of the black lipid membrane [27, 28]. Those early reports paved the way for the first report of liposomes and functional studies of the ionic transport across swollen phospholipids at the Babraham Laboratories in Cambridge [29]. Those major achievements by pioneering scientists including E.S. Johonnott, J. Folch, P. Mueller, A.D. Bangham, K.R. Blodgett, and I. Langmuir created a platform for the development of the understanding of transport processes in lipid bilayer membranes. Indeed, over the past two decades, there has been an explosion in the use of model membrane systems as a tool to better understand biological systems. There are numerous reviews of such model membrane systems [30–38]. Recent developments, however, have seen the use of model biomimetic membranes as more than an aid to better understand biological function and have now advanced to become a core engineered element in a new generation of biomimetic molecular devices. The challenge for any technology based on synthetic lipid membranes is device stability. Although challenging to engineer, good stability is usually achieved using lipid bilayers that are tethered to or supported by a substrate that may be nonporous (e.g., gold, silicon) or porous (e.g., polymer, gel), rather than using suspended lipid bilayers.

Neutron reflectivity is a powerful tool of physics to characterize the nanostructure of lipid bilayer and protein biomimetic membrane systems with a resolution of a fraction of nanometer along the normal depth direction. Using neutrons as probes provides the advantage of "gentle" interactions with the nuclei of the atoms, thus minimizing the risk of ionizing radiation damage for delicate biological materials. At the same time, the scattering contrast can be enhanced by the use of isotope substitution. This is particularly relevant for hydrogenous materials such as lipids and protein in water because replacing hydrogen with deuterium atoms has a dramatic effect on the contrast, but minimal effects on the samples. For systems immersed in liquid phases such as the biomimetic membranes described here, changing the isotopic composition of the liquid allows the scattering contribution from the aqueous components and those coming by other interfacial components to be disentangled.

Utilizing the neutron source of the Institut Laue Langevin (Grenoble, France), we performed a nanoscale characterization of tethered lipid bilayer membrane (tBLM) consisting of deuterated 1,2-dimyristoyl-sn-glycero-3-phosphocholine (d-DMPC) lipid bilayers tethered to a substrate through benzyl disulfide undecaethylene glycol phytanol (DLP) immobilized on the gold surface [39]. The areal density of the DLP tethering molecules was controlled by co-adsorbing spacer molecules of benzyl disulfide tetraethylene glycol (PSM). This architecture is comprised of a O11 length ethylene glycol reservoir and C20, half-bilayer spanning phytanyl tether lipids (DLP), chemically attached by a benzyl disulfide undecaethylene glycol phytanol to a gold substrate (Fig. 4.2).

The benzyl disulfide group provides a strong dispersive attraction to the gold, spaces out the ethylene glycol chains of the O11 glycol linkers to provide a greater water volume in the reservoir space, and prevents oxidization and polymerization of the sulfur groups when the compound is stored in solution. The DLP are scaffolding units onto which the lipid bilayer is formed and maintain it at a defined distance from the gold substrate. Co-adsorption of specific molar ratios of the DLP and polar spacer molecules (PSM) composed of benzyl disulfide TEG COOH is used to control the DLP areal density.

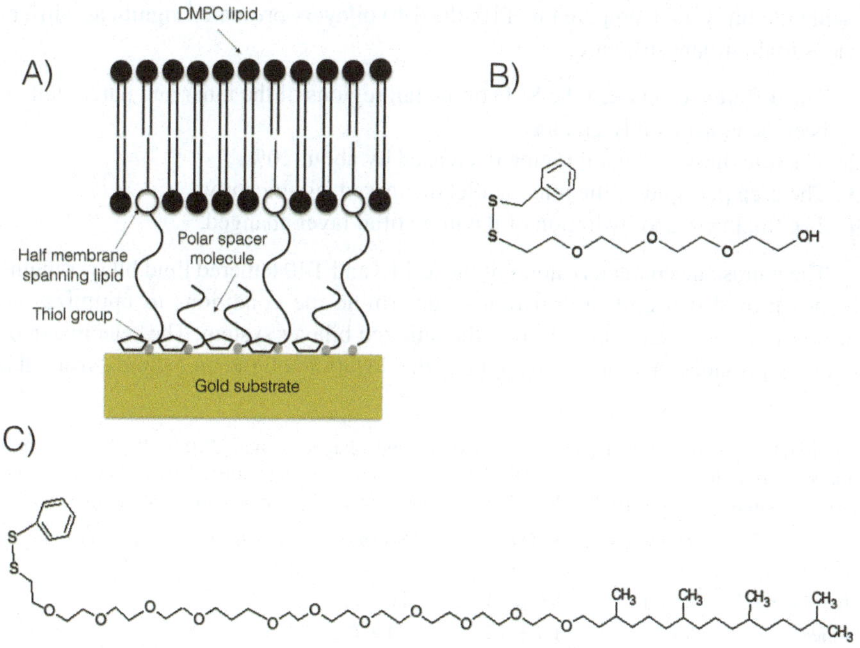

Fig. 4.2 (**a**) Sketch of the tethered lipid bilayer system. The half-membrane spanning lipids (empty heads) are immobilized on a gold substrate through polar linkers terminating with a sulfur group. Their areal density is controlled by coadsorption of polar spacer molecules. Free d-DMPC lipids with deuterated tails (solid heads) assembled together with the DLP complete the tethered lipid bilayer. (**b**) Chemical formula of the polar spacer molecule (PSM). (**c**) Chemical formula of the DLP. (Reproduced with permission from Ref. [39])

The tBLM illustrated in Fig. 4.2 was formed using an interlayer that comprised two different ratios of DLP:PSM, which were designated at T10 (10:90) and T40 (40:60). The nanostructural information we obtained using neutron reflectometry is useful to determine the conditions to optimize the insertion of membrane proteins into the tethered bilayer system. By using selective deuteration, we could distinguish the location of the free lipids from the tethered lipid and determine that the area density of tethers impacts both the distribution of the free and tethered lipids in the two leaflets and the properties of the interstitial layer between the lipid and the substrate [39]. The nanostructural parameters reported from those neutron reflectometry experiments is summarized in Tables 4.1 and 4.2, with an illustration of the nanostructural layers in Fig. 4.3. These nanostructural parameters provide a basis to consider the biological engineering of a lipid bilayer system that incorporates proteins.

The neutron reflectometry experiments quantified changes to the molecular structure of lipid bilayers tethered by T10 and T40 linkers corresponding to different molar ratios between DLP and PSM. The two leaflets of the bilayer shared the same features with the exception of the SLD of the lipid tail regions. The distinction between the isotopic composition of the d-DMPC free lipids and the DLPs provided the contrasts to distinguish differences in the molecular distribution of the two lipids inside the bilayers. Compared to T10, the T40 bilayers presented significant differences in their nanostructure:

1. The difference between the SLD of the tail regions of the inner and outer leaflets became considerably greater.
2. The thickness of the tail region decreased by about 20%.
3. The area per lipid of the inner leaflet decreased significantly.
4. The thickness and hydration of the interstitial layer changed.

The nanoscale characterization of those T10 and T40 tethered lipid bilayer membranes provided useful information to determine the conditions to optimize the insertion of membrane proteins into the tethered bilayer system. The selective deuteration provided a tool to distinguish the location of the free lipids from the

Table 4.1 Structural parameters of the tethered lipid bilayers system T10 at 20 °C for a lipid bilayer assembled from d-DMPC. The interlayer contained the ratio of DLP:PSM tethering molecules and spacers of 10:90. The values without error were fixed or kept at their nominal value

Layer	Thickness (Å)	SLD (× 10^{-6} Å^{-2})	Solvent (%)	Roughness (Å)	Area per lipid (Å^{-2})
Interlayer	11 ± 1	0.6 ± 0.1	26 ± 1	8 ± 1	–
Inner headgroup	8 ± 1	1.2 ± 0.2	34 ± 10	8 ± 1	43 ± 6
Inner tails	16 ± 1	3.9 ± 0.2	18 ± 8	8 ± 1	
Outer tails	16 ± 1	6.8	10 ± 5	8 ± 1	49 ± 3
Outer headgroup	8 ± 1	1.8	34 ± 10	8 ± 1	

Reproduced with permission from Ref. [39]

Table 4.2 Structural parameters of the tethered lipid bilayers system T40 at 20 °C for a lipid bilayer assembled from DMPC. The interlayer contained the ratio of DLP:PSM tethering molecules and spacers of 40:60. The values without error were fixed or kept at their nominal value

Layer	Thickness (Å)	SLD (× 10^{-6} Å^{-2})	Solvent (%)	Roughness (Å)	Area per lipid (Å^{-2})
Interlayer	17 ± 4	0.6 ± 0.1	64 ± 1	8 ± 0.3	–
Inner headgroup	7 ± 1	0.8 ± 0.2	19 ± 10	8 ± 0.3	44 ± 5
Inner tails	13 ± 1	0.9 ± 0.2	6 ± 5	8 ± 0.3	
Outer tails	13 ± 1	6.8	11 ± 5	8 ± 0.3	49 ± 3
Outer headgroup	9 ± 1	1.8	44 ± 10	8 ± 0.3	

Reproduced with permission from Ref. [39]

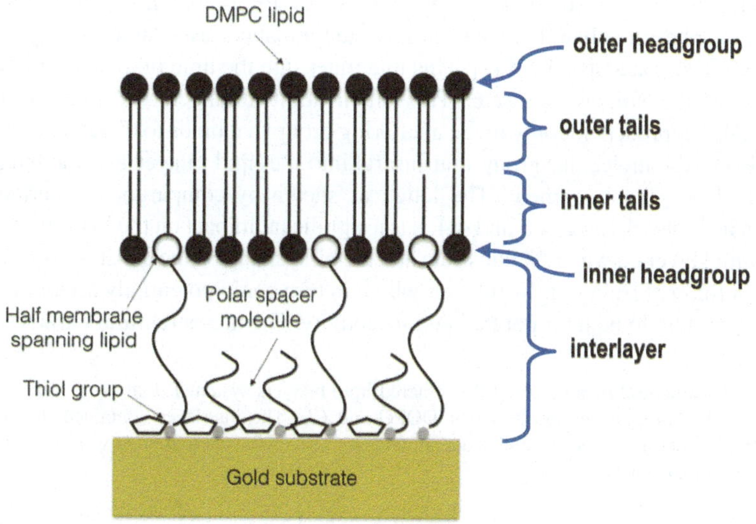

Fig. 4.3 Illustration of the nanostructural layers described in Tables 4.1 and 4.2. To obtain the nanostructural parameters, the interlayer was formed with two different ratios of PSM:DLP, which were T10 (10:90) and T40 (40:60)

tethered lipid and determine that the area density of tethers impacts both the distribution of the free and tethered lipids in the two leaflets and the properties of the interstitial layer between the lipid and the substrate.

A key to the development of lipid membrane-based devices is a fundamental understanding of how the molecular structure of the lipid bilayer membrane is influenced by the type of lipids used to build the membrane. This is particularly important when membrane proteins for ion transport are included in these devices since the precise lipid environment affects the ability to incorporate membrane proteins and their functionality. Lipid bilayer membranes of cells contain several types of lipids and other molecules, all of which can influence the incorporation of proteins that facilitate the transport of molecules and ions across the cell membrane. For example, the addition of cardiolipin (CL) to a membrane composed mainly of dioleoyl phosphatidylcholine (DOPC) caused distinct structural rearrangement including crowding of ethylene glycol groups of the tethering molecules in the inner head region and a thinning of the lipid tail region bilayer [40]. Those nanostructural effects were measured using lipid bilayer membranes tethered to surfaces using the T10 and T40 system (Fig. 4.2), and the nanostructural parameters are summarized in Tables 4.3 and 4.4.

An additional significant design consideration for the assembly of lipid bilayer biomimetic membranes is the type of surface and the chemistry used for tethering the lipid bilayer. Another approach to the design a tethering architecture included silanization of the silicon surface with (3-aminopropyl)triethoxysilane (APTES) molecules that produce covalent bonds with N-hydroxysuccinimide (NHS) terminated molecules that were a poly(ethylene glycol) (PEG) 2000 molecule bound to a distearoyl phosphoethanolamine (DSPE) lipid for that published tethering approach, which is designated as DPN [41]. This integrates into the lipid bilayer, providing an anchor to the silicon substrate. This silanization architecture differs from the PSM:DLP approach not only in the anchoring group that binds to different materials but also in the molecular group that inserts into the lipid bilayer and the length of the ethylene glycol segment. The latter, as shown by comparing the parameters shown in Tables 4.1, 4.2, 4.3 and 4.4, has a significant impact on the hydration of the interstitial layer between the substrate and the bilayer and on the constraints imposed on the proximal leaflet of the bilayer, which in turn could potentially affect the ability of a protein to be incorporated and its activity. The nanostructural parameters of

Table 4.3 Structural parameters of the tethered lipid bilayers system T10 at 20 °C for two lipid bilayers with different compositions of DOPC and CL. The interlayer contained the ratio of DLP:PSM tethering molecules and spacers of 10:90. The values without error were fixed or kept at their nominal value

Lipid composition	Layer	Thickness (Å)	SLD ($\times 10^{-6}$ Å^{-2})	Solvent (%)	Roughness (Å)
DOPC / CL (90:10)	Interlayer	15 ± 1	0.6	17 ± 3	9 ± 1
	Inner headgroup	6 ± 1	0.8 ± 0.4	16 ± 10	7 ± 1
	Inner tails	11 ± 1	−0.3	22 ± 2	7 ± 1
	Outer tails	11 ± 1	−0.3	22 ± 2	7 ± 1
	Outer headgroup	9 ± 1	1.9 ± 0.5	64 ± 7	7 ± 1
DOPC / CL (50:50)	Interlayer	11 ± 1	0.6	36 ± 9	7 ± 3
	Inner headgroup	18 ± 2	0.9 ± 0.2	10 ± 7	9 ± 2
	Inner tails	10 ± 1	−0.3	16 ± 4	9 ± 2
	Outer tails	10 ± 1	−0.3	16 ± 4	9 ± 2
	Outer headgroup	11 ± 2	1.9 ± 0.6	45 ± 10	9 ± 2

Reproduced with permission from Ref. [40]

Table 4.4 Structural parameters of the tethered lipid bilayers system T40 at 20 °C for three lipid bilayers with different compositions of DOPC and CL. The interlayer contained the ratio of DLP:PSM tethering molecules and spacers of 40:60. The values without error were fixed or kept at their nominal value

Lipid composition	Layer	Thickness (Å)	SLD ($\times 10^{-6}$ Å^{-2})	Solvent (%)	Roughness (Å)
DOPC / CL (90:10)	Interlayer	12 ± 2	0.6	27 ± 10	12 ± 2
	Inner headgroup	5 ± 2	1 ± 1	25 ± 20	7 ± 3
	Inner tails	12 ± 1	−0.3	22 ± 4	7 ± 3
	Outer tails	12 ± 1	−0.3	22 ± 4	7 ± 3
	Outer headgroup	11 ± 3	2.0 ± 0.8	45 ± 15	7 ± 3
DOPC / CL (50:50)	Interlayer	10 ± 1	0.6	25 ± 10	7 ± 2
	Inner headgroup	15 ± 1	1.0 ± 0.3	10 ± 4	7 ± 1
	Inner tails	10 ± 1	−0.3	26 ± 4	7 ± 1
	Outer tails	10 ± 1	−0.3	26 ± 4	7 ± 1
	Outer headgroup	12 ± 2	1.9 ± 0.5	41 ± 9	9 ± 1
DOPC / CL (20:80)	Interlayer	10 ± 1	0.6	28 ± 7	9 ± 2
	Inner headgroup	19 ± 1	0.6 ± 0.2	10 ± 7	7 ± 1
	Inner tails	10 ± 1	−0.3	28 ± 4	7 ± 1
	Outer tails	10 ± 1	−0.3	28 ± 4	7 ± 1
	Outer headgroup	12 ± 2	1.9 ± 0.4	50 ± 8	7 ± 1

Reproduced with permission from Ref. [40]

the lipid bilayer assembled with the DPN tethering system are summarized in Table 4.5.

Fundamental Nanostructure of a Tethered Lipid Bilayer Membrane with Incorporated Proteins

Indeed, the influence of protein incorporation was studied with the lipid bilayer systems summarized in Tables 4.4 and 4.5. The incorporation of NhaA in bilayers assembled using DOPC and CL utilizing the two different tethering protocols was quantified, and the way that protein incorporation modulated the nanostructural parameters of those lipid bilayer membranes is in Table 4.6 [40].

The incorporation of NhaA into the tethered lipid bilayer system that comprised DOPC/CL (50:50) and DLP/PSM (40:60) induced a remodeling of the lipid structure involving the thinning of the inner headgroup region, swelling of the lipid chain and outer headgroup regions, along with a significant reduction of the hydration in

Table 4.5 Structural parameters of the tethered lipid bilayers system DPN at 20 °C for three lipid bilayers with different compositions of DOPC and CL

Lipid composition	Layer	Thickness (Å)	SLD ($\times 10^{-6}$ Å^{-2})	Solvent (%)	Roughness (Å)
DOPC	Interlayer	18 ± 1	0.6	90 ± 3	7 ± 2
	Inner headgroup	11 ± 1	1.6 ± 0.2	14 ± 7	7 ± 1
	Inner tails	16 ± 1	−0.3	25 ± 2	7 ± 1
	Outer tails	16 ± 1	−0.3	25 ± 2	7 ± 1
	Outer headgroup	6 ± 1	1.88	36 ± 10	7 ± 1
DOPC/CL (50:50)	Interlayer	12 ± 1	0.6	70 ± 2	11 ± 1
	Inner headgroup	17 ± 1	1.6 ± 0.1	21 ± 3	7 ± 1
	Inner tails	12 ± 1	−0.3	23 ± 1	7 ± 1
	Outer tails	12 ± 1	−0.3	23 ± 1	7 ± 1
	Outer headgroup	6 ± 1	2.4 ± 0.2	27 ± 5	7 ± 1
DOPC/CL (80:20) *REGION 1 (surface coverage 83 ± 4%)*	Interlayer	49 ± 6	0.6	28 ± 7	9 ± 2
	Inner headgroup	20 ± 7	0.6 ± 0.2	10 ± 7	7 ± 1
	Inner tails	11 ± 3	−0.3	28 ± 4	7 ± 1
	Outer tails	11 ± 3	−0.3	28 ± 4	7 ± 1
	Outer headgroup	11 ± 7	1.9 ± 0.4	50 ± 8	7 ± 1
DOPC/CL (80:20) *REGION 2 (surface coverage 17 ± 4%)*	Interlayer	57 ± 4	0.4 ± 0.1	11 ± 11	3 ± 5

The interlayer contained the DSPE-PEG-NHS tethering molecules. The values without error were fixed or kept at their nominal value. Note that for the DOPC/CL 80:20 sample, a two-region model was adopted to fit the data. Region 1 comprises 83% of the surface covered with both the tethering molecules and the lipid bilayer. Region 2 comprises the remaining 17% of the surface that is covered only with the tethering molecules and no lipid bilayer. Reproduced with permission from Ref. [40]

the inner region of the lipid bilayer. The lipid chain regions underwent a drastic change in the SLD from a negative to a positive value. Normally the chain region has a low SLD (-0.3×10^{-6} Å^{-2}), because of the large number of hydrogen atoms in the hydrocarbon tail. Incorporation of a protein that has an SLD between 1.6 and 2.75×10^{-6} Å^{-2} (depending on the isotopic composition of the liquid phase) can significantly raise the tail region SLD, making it a very sensitive parameter to the degree of protein insertion. The SLD values of the lipid tail regions observed after the exposure to the protein indicate a protein volume fraction of 49 ± 5% inside the bilayer. We can suppose that the incorporation of the protein in such a significant amount induced a change of conformation in the lipid tails so that the thickness of this layer became commensurate with the size of the hydrophobic part of the

Table 4.6 Structural parameters of the two different tethered lipid bilayer systems with the transmembrane transport protein NhaA incorporated into each lipid bilayer

Tethered bilayer system	Layer	Thickness (Å)	SLD (× 10^{-6} Å^{-2})	Solvent (%)	Roughness (Å)
T40 (DLP/PSM of 40:60) and DOPC/CL (50:50) NhaA VF = 0.49 ± 0.05	Interlayer	11 ± 2	0.6	17 ± 3	9 ± 1
	Inner headgroup	4 ± 2	1.7 ± 1.2	16 ± 10	7 ± 1
	Inner tails	19 ± 1	1.2 ± 0.1	22 ± 2	7 ± 1
	Outer tails	19 ± 1	1.2 ± 0.1	22 ± 2	7 ± 1
	Outer headgroup	36 ± 5	2.3 ± 1.1	64 ± 7	7 ± 1
DSPE-PEG-NHS and DOPC/CL (50:50) NhaA VF = 0.19 ± 0.01	Interlayer	12 ± 2	0.6	85 ± 5	12 ± 3
	Inner headgroup	17 ± 1	1.8 ± 0.3	21 ± 3	7 ± 3
	Inner tails	13 ± 1	0.3 ± 0.1	17 ± 2	7 ± 3
	Outer tails	13 ± 1	0.3 ± 0.1	17 ± 2	7 ± 3
	Outer headgroup	7 ± 4	2.3 ± 1.5	71 ± 14	7 ± 3

The volume fraction (VF) of the NhaA protein in each bilayer system is indicated. The lipid bilayers has the same compositions of DOPC and CL. The interlayer contained either (i) the ratio of DLP/PSM tethering molecules and spacers of 40:60 or (ii) the DSPE-PEG-NHS tethering molecules. The values without error were fixed or kept at their nominal value. Reproduced with permission from Ref. [40]

protein. The length of the hydrophobic zone of the NhaA protein along its axis is around 35 Å, a value compatible with the total thickness of the hydrophobic chain region obtained from the fits following the incorporation. The lipid chain regions experienced a significant reduction in the water content. It is reasonable to assume that in the course of the incorporation process, the protein may have replaced some lipids of the tBLM but also used the space provided by existing defects in the membrane. Either way, the lower solvent content in the chain regions after protein incorporation suggests that the incorporation may be accompanied by a repairing effect on the bilayer. This is also plausible because the protein transfer from the liposomes could include transfer of lipids from the liposome to the bilayer that may fill preexisting defects.

For the tethered lipid bilayer system that comprised DOPC/CL (50:50) and DLP/PSM (40:60), the presence of the NhaA protein in the lipid headgroup regions affected the inner and the outer leaflets differently. The decrease in thickness of the inner headgroup region could be caused by the presence of NhaA displacing part of the CL lipids. If the presence of CL promotes the crowding of ethylene glycol units in the headgroup region as shown in the previous section, a reduced amount of this lipid might explain the thinning of this layer. The presence of the protein might also impose a degree of order to the structure of the bilayer, further contributing to the expulsion of ethylene glycol units. The decrease in solvent content of the submembrane space is also consistent with the displacement of ethylene glycol units that were previously accumulated in the inner headgroup toward the submembrane space. The increase of the effective thickness of the outer headgroup layer might be instead related to the histidine tag, which is a remnant of the protein purification. This tag is assumed to protrude from the membrane or lay on top of the bilayer surface. Even though the orientation of NhaA is not controlled in the incorporation process, the presence of such a bulky molecule on one side of the protein could induce a preferential orientation of the protein that would insert into the lipid bilayer with the histidine group pointing in the outward direction. The opposite orientation would be less favorable due to the lack of space in the submembrane region crowded by the tethering molecules.

The tethered lipid bilayer system that comprised DOPC/CL (50:50) and DSPE-PEG-NHS allowed a much lower amount of NhaA protein to be incorporated. Even with the lower amount of protein incorporation, there was an increase of the SLD of the inner chains from a negative to a positive value following the incorporation of the NhaA. The NhaA incorporation also induces a thickening of the hydrophobic tail region and a reduction of the water content, but to a smaller extent, from 24 to 26 Å and from 23% to 17%, respectively.

For biological engineering and applications to nanomedicine, the importance of these detailed nanostructural analyses is the information for better understanding and hence control of the design and function of lipid bilayer biomimetic systems. The modulation of the thickness of the lipid bilayers following the NhaA incorporation is related to the so-called hydrophobic mismatch, which originates from the difference between the thickness of the hydrophobic regions of the lipid bilayer and the protein that spans across the membrane [42]. This process affects significantly the organization of lipids and proteins, but the exact implications for the structure and function strongly depend on the individual protein and the lipid system. It can induce different structural rearrangements including protein oligomerization, protein and lipid conformational changes, and membrane curvature. For example, for tethered membrane systems (cf. Figs. 4.1 and 4.2), the difference in the tail region thickness before and after the incorporation of the protein is significant. Among the reasons that might cause this difference is the low number of lipid species in the bilayer, which might limit the conformational ability of the bilayer to adapt to the hydrophobic domains of NhaA. Since the hydrophobic mismatch has a direct effect on the protein functionality, it is important for the development of devices based on lipid protein systems to assess the activity of the incorporated protein as a function of the lipid mixture, in view of the reduced number of lipids normally available in artificial membranes with respect to real biological ones.

Similar nanostructural changes in lipid bilayers are induced by the incorporation of other transmembrane ion-conducting proteins. For example, the porin OprF has a central role in *Pseudomonas aeruginosa* virulence and thus provides a putative target for either vaccines or antibiotic cofactors that could overcome the bacterium's natural resistance to antibiotics. Biomimetic lipid membrane systems are important to investigate nanostructural aspects and function of OprF using neutron reflectometry and electrical impedance spectroscopy. The pore-forming functions of OprF proteins are measured from biomimetic lipid bilayer membrane systems using impedance spectroscopy. Table 4.7 shows the nanostructural parameters of the

Table 4.7 Structural parameters of a tethered lipid bilayer system T10 at 20 °C before and after the incorporation of the transmembrane porin OprF. The lipid bilayer used for the T10 systems was assembled from non-deuterated DMPC. The interlayer contained the ratio of DLP:PSM tethering molecules and spacers of 10:90

Tethered bilayer system	Layer	Thickness (Å)	SLD ($\times 10^{-6}$ Å^{-2})	Solvent (%)	Roughness (Å)
T10 and DMPC before OprF	Interlayer	17 ± 2	0.6 ± 0.1	23 ± 7	8 ± 5
	Inner headgroup	8 ± 2	1.2	66 ± 20	11 ± 5
	Inner tails	13 ± 2	−0.35	40 ± 5	11 ± 5
	Outer tails	13 ± 2	−0.37	40 ± 5	11 ± 5
	Outer headgroup	8 ± 2	1.8	64 ± 20	11 ± 5
T10 and DMPC after OprF	Interlayer 1	10 ± 2	0.6	34 ± 10	6 ± 5
	Interlayer 2	12 ± 2	2.1 (1.4) ± 0.3	8 ± 10	7 ± 5
	Inner headgroup	9 ± 2	2.3 (1.4) ± 0.3	52 ± 20	7 ± 5
	Inner tails	10 ± 2	1.7 (0.9) ± 0.1	31 ± 9	7 ± 5
	Outer tails	10 ± 2	1.7 (0.9) ± 0.1	31 ± 9	7 ± 5
	Outer headgroup	10 ± 2	2.7 (1.9) ± 1.5	58 ± 20	7 ± 5
	Extra-membranous	20 ± 5	2.3 (2.0)	77 ± 10	9 ± 5

Note that in the layers in which the OprF is present, two values of SLD are listed due to the presence of labile hydrogen atoms that exchange with deuterium. The first value corresponds to the D_2O contrast, whereas the value in parentheses corresponds to the H_2O contrast. The extra membranous layer is due to the presence of the C-terminus of OprF that protrudes outside the membrane. The values without error were fixed or kept at their nominal value. Reproduced with permission from Ref. [43]

incorporation of OprF into biomimetic lipid bilayer membrane systems were measured using neutron reflectometry and a T10 tethered lipid membrane system assembled from protonated DMPC and DLP/PSM (10:90) [43]. The tethered system is similar to the one described in Table 4.1; except that in Table 4.7, the DMPC lipids are not deuterated.

Neutron reflectometry gave a precise nanostructural characterization of the tethered bilayer system and identified differences occurring before and after the incorporation of OprF. The reflectivity profiles correspond to OprF incorporated into the DMPC bilayer with a volume fraction of 0.56 ± 0.03 (56% of the bilayer volume). The precise modeling of the reflectivity data confirms that OprF is incorporated in the lipid bilayer with extra membranous parts above and below the lipid bilayer. Although the model is based on the most likely conformation of the incorporated OprF (the closed state conformer) with the C-terminus outside the membrane, we cannot exclude that some of the proteins are in the open conformation. In the open conformation, although the C-terminus is thought to become embedded in the intra-membranous region, extra-membranous loops are still present on both sides of the membrane. Furthermore, the structural information obtained by neutron reflectometry is averaged from measurements that take hours to complete, during which the conformation of the porin fluctuates between the open and closed conformers. However, the nanostructural parameters in Table 4.7 show that the protein is more extended on the side of the membrane pointing toward the solvent (~20 Å) compared to the side pointing toward the substrate (~12 Å). This might suggest a preferred orientation of the protein with the voluminous C-terminus preferentially oriented toward the solvent. For example, this orientation might be favored due to the larger space available in the direction of the solvent compared to the side closer to the substrate, where the C-terminus would compete for space with the tethered molecules. The incorporation of the OprF is linked also to a reduced hydration of the lipid bilayer, since the incorporation of the protein is accompanied by the displacement some water molecules from the lipid bilayer. In this regard, it is important to note a tethered system formed by the DMPC is characterized by a low hydrated interlayer and a partially hydrated tail region.

In considering the development of a drug delivery system based on liposomes, it is worth exploring the role that neutron scattering can play in measuring the nanostructure of liposomes that incorporate proteins. A detailed structural characterization of liposomes containing molecular drugs or therapeutical agents is a key factor for their development and optimization. Size and lipid composition in addition to the presence, amount, localization, and orientation of nonlipidic molecular components embedded in the lipidic membrane may influence the interaction of liposomes with human cells. The use of small-angle neutron and X-ray scattering provides averaged structural information on length scales ranging from micrometer to subnanometer scales. When applied to liposomes, these scattering techniques provide information about size and shape as well as a variety of details of the lipid and of the nonlipid components, such as thickness, roughness, and hydration of the head and tail regions, amount, and localization of guest molecules [44].

That system of nanostructural analysis was applied to the incorporation of OprF into liposomes, since OprF is the main porin of the Gram-negative bacterium *Pseudomonas aeruginosa* that is responsible for a significant number of nosocomially acquired infections that are resistant to most of the common antibiotics. OprF is also involved in the mechanisms of adhesion with the host cell and in the interaction with γ interferon for evasion of the host immune system. It was also recognized that OprF might impact the biogenesis of outer membrane vesicles used for cell-to-cell communication and for trafficking hydrophobic signals within a bacterial population. The system of nanostructural analysis is easily adapted to different classes of liposome systems containing different host molecules in the bilayer, in the interior, or on the surface of the liposome, providing a useful tool to obtain a detailed average structural characterization that could be difficult to obtain with other techniques. The analysis provided detailed information on the lipid bilayer structure, the amount of OprF proteins in the lipid bilayer, their average localization, orientation, and the effect of the protein incorporation on the lipid bilayer. This information is valuable for fine-tuning the proteoliposome delivery system, for example, in quantifying the dosage of proteins delivered for therapy [44].

Neutron reflectometry provides a powerful tool for the measurement of nanostructural parameters of lipid/proteins systems where the system is assembled using cell-free purification procedures. Cell-free expression in the presence of a supported membrane has been used to study membrane-protein interactions but has been unable to provide information about protein conformation or structure. Soranzo et al. [45] coupled neutron reflectometry with cell-free expression and demonstrated the expression of membrane protein (p7) directly into lipid bilayers and obtained details of the protein structure and conformation with sub-nanometer resolution [45]. The structural data obtained by neutron reflectometry are the first data reported for the p7 protein genotype 1a (H77 strain) and the first in a biologically relevant lipid environment. Further, the use of deuterated amino acids in the synthesis of the protein enhances the contrast of the scattered neutrons and unambiguously allows the differentiation between the incorporated protein and the (protonated) lipid bilayer. This approach can be further developed by expressing proteins where only specific amino acids are isotopically labeled and performing a series of neutron measurements, using either reflectivity or small angle scattering, to more precisely map out protein structures and conformations.

Concepts of Biocompatibility to Protect Implanted Devices to Function for Long Periods Inside the Body

We can interpret "stealth" protection as providing some means for an implanted device or nanoparticle to avoid becoming a target of the immune system. Such "stealth" protection is crucial if the implanted device or nanoparticle needs to function properly for long periods inside the body and in-contact with body fluids and tissues. This requires that the implanted devices do not induce long-term inflammation and also invoke the concept of "biocompatibility" between the body and an implanted (bio)material.

Scientists such as B. Ratner, R. Langer, J. M. Anderson, and D. Williams pioneered the history of modern biomaterials and biocompatibility issues [46–52]. Although the literature has included the keyword "biomaterial" since the second world war (e.g., in PubMed search terms), the term "biomaterial" had been used in several different contexts as a material derived from biological processes [53] or simply as describing a sample of body tissues [54]. It is only recently that a coherent definition of "biocompatibility" has been described in the literature. For example, in 1976, the first definition describes a biomaterial as "a nonviable material used in a medical device, intended to interact with biological systems" [50]. This definition has evolved gradually over more than 40 years. Every decade has brought steps toward precision in the definition, following the emergence of new materials and concepts. In 2009, Williams expanded the definition taking into account nanoparticles, hydrogels, and self-assembly and structuration of biological systems [55]. That definition of Williams focused on the usage (the final biological goal), rather than on the biomaterial itself. In other words, the biomaterial was no longer considered as inert and the word "material" has been replaced by "substance" to highlight the diversity of biocompatible structures. The most recent definition though by

Williams, in 2014, quoted that "A biomaterial is a substance that has been engineered to take a form which, alone or as part of a complex system, is used to direct, by control of interactions with components of living systems, the course of any therapeutic or diagnostic procedure, in human or veterinary medicine" [56]. The significant change in conception has been to consider that biocompatibility is the result of a biomaterial interacting with the body as a system. The focus of the biocompatibility pathways for a biomaterial has been the way in which the components of the biomaterial actually interact with cells of the body [57]. As is summarized in Fig. 4.4, Williams [56] has identified those components acting on the cell as having an origin from the biomaterial (e.g., metal ions, polymer additives, nanoparticles, contaminants, etc.), mechanical and biophysical mediators of interactions (e.g., electromagnetic fields, structural/fluid stresses), internalization mechanisms (e.g., phagocytosis, endocytosis; pinocytosis), or material mediators of interactions (chemical structure, elasticity, shape, volume, topography). However, those mediating components remain as originating from, or derived from, the biomaterial. Nonetheless, such a concept of biocompatibility subsumes a collection of individual phenomena that make it impossible to consider a scale of biocompatibility. Furthermore, even by referring to a "biocompatible system" rather than a "biocompatible material," Williams concluded that the term "biocompatible system" remained imprecise and potentially misleading [56].

Such imprecision can be avoided if a biocompatible system is defined by the intended functional properties of that system. Such a definition would consider that

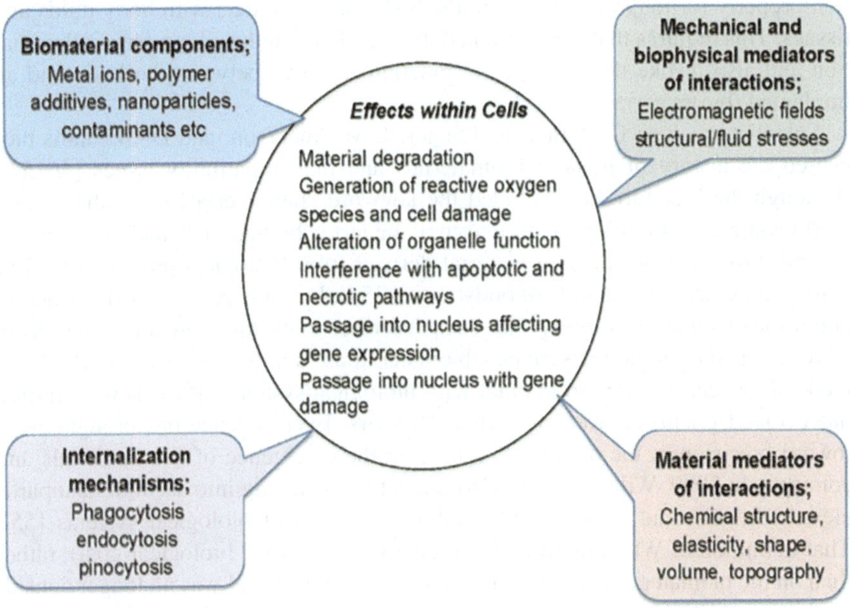

Fig. 4.4 Summary of mediators of biocompatibility and critical cellular. (Reproduced with permission [56]. Copyright 2014, Elsevier)

the implanted device would need the characteristics of communication with the body in order to be a symbiotic device ("symbio-bot"). The value of re-defining the framework of a "biocompatible system" is in considering the next generation of such symbiotic devices where long-term function is the focus of the implanted system. Currently, implanted medical devices that are designed to restore a function or enhance body reconstruction look fine but remain foreign devices for the body and are not fully biocompatible. This is due to interfacial interactions, with mechanotransductory and sterile inflammatory mechanisms involved as essential and primary mechanisms [57]. The body response to the implantation of medical devices is largely well documented in the literature [58–62]. The first step in this response appears within seconds after implantation and is a protein coating followed by cell adherence. This is the first stage of biofouling that is the target for most of the research in this field [63–65]. After this early inflammatory phase, which can result in the implant rejection, a foreign body response (FBR) occurs. Much research has been directed at ways to avoid such a chronic inflammation that leads to encapsulation of the device, especially targeting ways to control M1-to-M2 macrophage differentiation [66, 70].

When macrophages cannot degrade the implanted object, they fuse and form polynuclear giant cells [71]. Those giant cells will either degrade the implant or create a cystic capsule. A porous implant will contain numerous macrophages and giant cells while smooth surface will produce a fibrin coating [71–73]. The nature and the level of the inflammatory response depends also upon duration and location of the implantation. For example, the lifetime of breast implants is limited to 10 years [74]. In the cardiovascular system, thrombus formation limits the lifetime of implanted devices such as stents [75]. An approach such as grafting PTFE grafted on PLA for stents can improve blood biocompatibility by allowing albumin to passivate stent surface, with the benefit that platelets are less adherent and activated [76].

The immune system is highly effective in coating any foreign particle or surface with a biomolecular corona, particularly when the foreign particle or surface contacts blood [77]. There have been many different types of surface treatments published, each with varying degrees of efficacy in reducing coating and hence improving the "stealth" characteristics of the implanted device or nanoparticle [78].

A physically absorbed monolayer of 1-stearoly-2-hydroxy-sn-glycero-3-phosphocholine (PC) has been shown to resist protein adsorption but had low stability when exposed to Triton X-100 surfactant and ethylene oxide sterilization. The surfaces utilized in those experiments were silicon wafers and glass slides that were silanized using 3-(trimethoxysilyl) propyl methacrylate (TSM). However, grafting the PC to the surfaces using in situ polymerization (poly-PC) greatly improved the stability of the poly-PC monolayer when exposed to Triton X-100 and ethylene oxide sterilization. This poly-PC monolayer greatly reduced macrophage adhesion and the formation of foreign body giant cells when tested in vivo [79].

This concept to create a bio-interface to control cell interactions was applied to metallic surfaces coated with four types of phospholipids, which included 1-pal mitoyl-2-oleoyl-*sn*-glycero-3-phosphocholine (POPC), 1-palmitoyl-2-oleoyl-*sn*-glycero-3-phosphoethanolamine (POPE), 1-palmitoyl-2-oleoyl-*sn*-glycero-

3-[phospho-L-serine] (POPS), and 1-palmitoyl-2-oleoyl-*sn*glycero-3-phospho-*rac*-(1-glycerol) (POPG). Those phospholipids were assessed in their ability to "hide" the metallic surfaces, which were either mirror-polished pure nonporous titanium, mirror-polished porous Ti6A17Nb titanium, or vacuum plasma sprayed titanium. The interpretation of the results from those experiments was difficult because the lipids were applied to the metallic surfaces using a very simple evaporation process that did not produce uniform and thin membranes, but rather loosely bound and very rough multilamellar stacks on the metallic surfaces. For example, a 2-hour exposure to phosphate-buffered saline showed the lipid coating was not very homogeneous, as imaged by scanning electron microscopy. It was very likely that the inhomogeneity of the surface-adsorbed phospholipids led to removal of adhering cells during exchanges of culture medium. Generally, the phospholipid coatings led to a reduced viability of adherent macrophages, which led to the simple conclusion that these experiments supported the usage of the phospholipids POPC and POPE to improve the performance of metallic implants [80].

This concept was utilized in an attempt to improve bone mineralization on titanium alloys. The aim was to mimic bone mineralization by cell matrix vesicles by adding a coating of calcium-binding phospholipids, such as phosphatidylserine (PS) and phosphatidylinositol (PI). Although PS and PI did increase the rate of bone mineralization, the phospholipids were applied in a simple procedure that resulted in 100 μm thick coatings. These thick coatings were not mechanically stable, leading the authors to conclude that thinner coatings need to be applied to ensure the mechanical stability of the implant [81].

Those previous publications indicated that the biocompatibility of an implant is improved with a coating of phospholipids, but the key issue is to ensure that the phospholipids assemble into a thin layer that is anchored to the implant to provide a mechanically stable membrane to mimic the natural cell membrane. To stabilize the phospholipids onto a surface and to provide the conditions for the assembly of a membrane to mimic the cell membrane, then it is extremely important to develop a suitable anchoring layer of molecules. This anchoring layer of molecules would provide an interface between the surface of the implant and the assembly of a phospholipid membrane, so as to provide a long-term phospholipid membrane that cannot be simply dislodged by the movement of external solutions nor adhering cells.

A Lipid Bilayer Membrane Could Provide an Ideal Interface Between the Body and an Implanted Device

Since a lipid bilayer membrane is the basic external structure for all cells, it would seem rational to consider that the long-term performance of an implanted medical device would benefit considerably by adding a coating of a lipid bilayer membrane. Particularly since the majority of medical devices are constructed from materials that are not part of the body. This inspiration is from the healthy endothelial cell membrane, which remains a quiescent and stable lining for blood vessels. Indeed, Meri [82] describes one of the very fundamental processes in biology and a property of the immune system is to distinguish structures of "self" from "nonself." The alternative pathway (AP) of complement can recognize a broad spectrum of nonself structures by default. Anything not actively recognized as "self" is considered as "nonself" and will be labelled for destruction, for example, by opsonophagocytosis or direct killing. There is a complex cascade of complement factors that bind to target surfaces to active the AP. Similarly complex is the soluble factor H-mediated protection of cells against complement. In considering the problem of "self" versus "nonself" for implanted medical devices, an AP regulatory mechanism that can be exploited by an added lipid bilayer is to incorporate regulating membrane-bound proteins that operate at the C3 step of AP that initiates the C3b opsonization of the target surface and formation of membrane attack complexes (MAC). These membrane-bound proteins include complement receptor type 1 (CR1, CD35), membrane cofactor protein (MCP, CD46), decay-accelerating factor (DAF, CD55), and protectin (CD59), which are attached to cell membrane phospholipids via a glycophosphoinositol-linkage [82].

In pursuing this line of thought for a lipid bilayer interface as a means to improve the long-term performance of implanted medical devices, we need also to consider several other concepts derived from the function of the natural immune system that have been highlighted by Meri [82]. Malignant tumor cells and pathogenic microbes cause different kinds of problems, but common to both is that they usually manage to escape complement attack. Thus, they resemble "self." For tumor cells, this is natural, because most molecules that they carry on the surfaces are present also on normal nonmalignant cells. Some tumors have additional mechanisms to protect themselves against complement attack, like secretion of complement inhibitors or production of additional protective layers of carbohydrates (like mucins) on their surfaces [82].

To return to the inspiration from healthy endothelial cells in seeking an interface that will render an implanted medical device as "self." The healthy endothelial cell membrane resists the binding of factors from the complement cascade due to the natural secretion of an extacellular glycocalyx and expression of the membrane receptor CD59 that acts as an immune regulator by preventing pore formation by the complement membrane attack complex (MAC) [83]. Other membrane-associated proteins such as membrane cofactor protein (MCP/CH4) and decay-accelerating factor (DAF/CD55) also assist the regulatory mechanisms that protect host cells for complement damage [84]. The vascular endothelial glycocalyx, which comprises heparan sulfate and proteoglycans such as syndecain-1, has a critical role to prevent the adhesion of inflammatory cells onto the endothelial surface [85]. Damage to the glycocalyx allows access for complement factors to bind to phospholipid headgroups, particularly, for example, to anionic phospholipids and sphingolipids including phosphatidic acid, phosphatidylserine (PS), phosphatidylinositol phosphate (PIP), phosphatidylinositol 4,5-bisphosphate (PIP_2), phosphatidylinositol (3,4,5)-trisphosphate (PIP_3), cardiolipin (CL), sulfatide (ST), ceramide-1-phosphate (Cer1P), and dihydroceramide-1-phosphate (DHCer1P) [86]. An earlier report indicated that classical complement pathways are activated by complement factors, such as C1q or Factor H, binding to anionic phospholipids including cardiolipin, phosphatidylglycerol (PG), phosphatidylinositol (PI), and phosphatidylserine (PS) [87].

At this stage, we can form the hypothesis that a lipid bilayer membrane could function as a "stealth" membrane interface to render an implanted medical device as "self" to minimize excessive complement activation for the immune system by incorporating at least one membrane-bound protein of CD35, CD46, CD55, or CD59. Additionally, the lipid bilayer membrane would be further coated by a layer of molecules to mimic the natural glycocalyx.

Adding a Stable Lipid Bilayer Membrane to an Implanted Medical Device

Several approaches have been published to provide this stabilizing layer of anchoring molecules. We discuss these approaches here and utilize the term of tethered bilayer lipid membrane (tBLM) to describe the mechanically stable and planar phospholipid bilayer that is supported by an array of tethering molecules [88, 89]. These are often composed of a hydrophilic polymer whose extremities end with a group that anchors to a molecularly smooth substrate and a group that inserts into the lipid membrane.

The tethering molecules serve as a framework onto which the lipid bilayer assembles. In this respect, tBLMs show similarities with biological membranes characterized by the presence of cytoskeletal proteins that provide interaction and anchor points similar to the membrane tethers [90]. In addition to providing good stability [90–94], they offer a physical separation from the substrate that can accommodate extra-membranous segments of membrane proteins incorporated in the bilayer, as well as a hydration reservoir between the bilayer and the substrate. This space enables enhanced mechanical flexibility, in contrast to solid-supported lipid bilayers [88, 94, 95]. Hence, tBLMs possess some of the fundamental properties of biological membranes, while also offering a simplified platform to develop lipid-based bio-inspired devices.

Various tethering architectures and ligands have been reported to bind the tBLM to different substrates. Cornell et al. were one of the first groups to incorporate a tBLM in a functional device [88]. The building blocks of that tBLM were half-membrane-spanning phytanyl lipids tethered on gold substrates coadsorbed with polar spacer molecules composed of benzyl disulfide tetraethylene glycol to control the areal density of the tethered lipids [96, 97]. Variants of that tBLM with a double phytanoyl chain bound to the spacer [98, 99] or with different spacer molecules [100] were also developed to enhance the flexibility of the properties of the final tBLM.

Other tBLM architectures include synthetic 1-thiahexa-(ethylene oxide) lipids in combination with β-mercaptoethanol [101], or 1,2-dipalmityl-3-[ω-mercaptononа(ethylene oxide)] glycerol [102], or archaeal tetraether lipids [103]. The possibility to build a tBLM from affordable and abundant components such as commercial surfactants has been explored [104]. This could avoid the synthesis of complex organic molecules often needed to build tBLM and hence simplify the large-scale development of tBLM-based commercial devices. tBLMs mimicking microbial outer membranes were obtained using lipopolysaccharides extracted from *Escherichia coli* strains [105] with increased structural stability. tBLM with terminal deuterated anchors were also developed to exploit techniques sensitive to isotopic variation [106]. More recently, novel peptide tethered lipid bilayers (peptBLMs) [107, 108] were developed in which the tethering molecules are made up of peptides. The advantage of using peptides as tethering molecules is that their length, secondary structure, and hydrophobicity can be modulated by an appropriate choice of the amino acid sequence. All of the previous examples concern tBLMs adsorbed on gold substrates, but by changing the binding molecule, it is possible to adapt the tBLM architecture to different substrates such as glass [109] or silica [110–113].

The major advantage of a stable tBLM is to "hide" the implanted device from the inflammatory and immune responses of the body. The tBLM must provide a feature that can overcome the opsonization that habitually occurs for all implanted devices. Hence, the tBLM assists the implanted device to avoid becoming a target for the immune system, which requires that the implanted device or nanoparticle avoids the coating of either a collagenous capsule when implanted in body tissues [114] or proteins/opsins when in contact with blood [115]. As a starting point to consider the value of utilizing a stable tBLM to avoid opsonization, we can ask the question whether liposomes resist the processes of opsonization. Even if previous published research has not assembled liposomes from the precise composition of phospholipid membranes in the body, we can gain insights from previous research into the interaction of liposomes with body fluids. It has been known for several years that complement opsonization is an important factor for the phagocytic response of the body to remove liposomes, comprising negatively charged phospholipids, from the circulation. The negatively charged phospholipids included phosphatidylserine (PS), phosphatidylinositol (PI) and phosphatidic acid (PA) [13], or phosphatidylglycerol (PG) in different ratios with DMPC depending on the degree of negative charge desired in the liposomes [116].

It is also useful to consider that simply engineering the surfaces of liposomes or nanomedicines with PEG-PL coatings may not be the complete solution to avoid activating the immune system [117]. PEG–phospholipid conjugates exhibit low critical micelle concentration values and form stable micellar structures of approximately 30 nm in size, allowing drug solubilization, and are amenable for additional functionalization and parenteral administration. The application of PEG–phospholipid conjugates in steric stabilization and generation of long circulating oil-in-water nanoemulsions is also noteworthy. Finally, apart from nanomedicine design and development, PEG–phospholipids can further stabilize entities such as carbon nanotubes through surface adsorption and enhance their dispersion in aqueous media; such modifications also affect nanotube pharmacokinetics following intravenous injection for biological and mechanistic studies. There are clinical reports of acute hypersensitivity reactions to infusion of long circulating regulatory-approved PEGylated liposomes in sensitive individuals. These reactions, classified as pseudoallergy, are often associated with flushing and circulatory disturbances. Recent studies in pigs have suggested that the cardiopulmonary distress caused by Doxil® and other liposomes strongly correlates with complement activation. Accordingly, complement activation by PEGylated vesicles is an unexpected phenomenon since surface mPEG coverage is generally believed to dramatically suppress particle–protein interaction and blood opsonization events, including complement fixation. Clinical evidence from PEGylated liposome administration clearly attests that complement activation per se cannot solely explain vesicle-induced pseudoallergy. The involvement of other biological factors in pseudoallergy must not be ignored and should be investigated. Among many factors, there may be genomic differences between sensitive and nonsensitive individuals and future immunogenomic studies are expected to have an impact in pseudoallergy research. However, parallel development of in vitro tests is still needed to differentiate between immunologically (at least from complement immunology point of view) "potent" and "safe" nanomedicines and/or sensitive and nonresponding individuals. This is especially relevant to clinical scenarios where cardiovascular stress is not acceptable [117].

Lipid Bilayer Coatings to Enhance "Stealth" Protection for Nanoparticles

It would seem intuitively obvious to consider that a lipid coating could improve the "stealth" characteristics of an implanted medical device or nanoparticle. This would be especially so if the lipid coating could be a good mimic of the natural lipid membranes of cells in the body. A major problem to overcome is the protein corona that is established around all nanoparticles. There are a very large number of different proteins adsorbed to form the corona. For example, there are at least 162 different corona proteins associated with DOPG and silicon nanoparticles [116]. To overcome or to control the protein corona, there are a large number of potential lipids, in a large number of combinations and concentrations, to utilize in coatings. Perhaps the problem is not to try and identify the "one" type of coating, but to identify the combinations of lipids for the coating that minimizes the adsorption of the large possible number of corona proteins. This would inevitably need as systems approach to find the appropriate solution.

An excellent starting point for finding such a systems-based solution is the extensive literature review published recently by Hajipour et al. [118]. This systems approach to find the most appropriate lipid coating does not preclude other methods to minimize the adsorption of a protein corona, such as including PEG or modulating the shape of the nanoparticles to influence the adsorption of protein corona.

Most druggable cellular targets are membrane components, including membrane proteins and soluble proteins that interact with ligands or receptors embedded in membranes. Current target-based screening and intermolecular interaction assays generally do not include the lipid membrane environment in presenting these targets, possibly altering their native structure, and leading to misleading or incorrect results. To address this issue, an ideal assay involving membrane components would (i) mimic the natural membrane environment, (ii) be amenable to high-throughput implementation, and (iii) be easily multiplexed [118]. A lipid bilayer membrane can be used coat nanoparticles to improve their functions as diagnostic, drug delivery, or therapeutic systems. Lipid vesicles spontaneously fuse and assemble into a lipid bilayer on planar or spherical silica surfaces and other substrates. These supported lipid bilayers (SLBs) maintain characteristics of biological membranes and are thus considered to be biomembrane mimetic systems that are stable because of the underlying substrate. Examples of their shared characteristics with biomembranes include lateral fluidity, barrier formation to ions and molecules, and their ability to incorporate membrane proteins into them. Biomimetic silica microspheres consisting of SLBs on solid or porous silica microspheres have been utilized for different biosensing applications. The advantages of such biomimetic microspheres for biosensing include their increased surface area to volume ratio which improves the detection limits of analytes and their amenability for miniaturization, multiplexing, and high throughput screening [119]. Biomimetic nanoparticles can be prepared through direct "biological modeling" or "bio-mixing" methods. Biomimetic nanoparticles developed from protein/peptide templates utilize the characteristics of proteins or peptides to bind metal ions through different metal binding sites, such as *N*-terminal amines or multi-metal binding sites, thus providing templates for nanoparticle growth. Biomimetic nanoparticles developed by coating with cell membranes are basically a core-shell that incorporate different cell types such as leukocytes, platelets, and red blood cells into with synthetic nanoparticles (e.g., polymeric NPs, Au, and silica) [120]. The use of biomimetic nanoparticles offers many distinct advantages, including enhanced stability, the solubilization of hydrophobic payloads, extended blood residence times, and the ability to better target a region of interest.

A novel type of biomimetic nanovesicle (BNVs) has been produced from the cytoplasmic membranes of mesenchymal stem cells (MSCs) and termed Nano-Ghosts (NGs) [121]. The MSCs undergo chemical and physical process that enables the exclusion of their cytoplasmic content while maintaining an intact membrane with the correct orientation. MSCs are treated with a hypotonic buffer followed by homogenization and serial centrifuges for their cytoplasmic removal, obtaining cell Ghosts. The Ghosts are either extruded or sonicated to downsize them and achieving NGs. NGs are further cleaned from any residual by centrifugation. The NGs pellet is collected via ultracentrifugation and resuspended in the buffer of choice. PEGylation is performed on the final NGs product to further enhance their stability in solution. The NGs retain the composition, orientation, and function of the MSCs membrane in an inanimate particle. These characteristics are particularly the capability to target and infiltrate malignant tissues while rapidly clearing from other organs with no reported off target effects. The utility of the NGs in selectively delivering diverse therapeutics, ranging from large proteins to nucleic acids, was demonstrated with various tumor models, among them pancreatic, lung, and prostate cancer, presenting encouraging therapeutic potential. Moreover, the in vivo homing of the NGs can be modulated by culturing the MSCs source cells in the presence of either cytokines or cancer-derived conditioned media [122]. These are important characteristics of the NGs, like MSCs, to provide the possibility to target therapies toward inflammatory eases.

An interesting observation is that almost all particles with lipid coatings have been spherical despite emerging evidence that nonspherical shapes can provide important advantages including reduced nonspecific elimination and increased target-specific binding. Indeed, ellipsoidal particles have been shown to provide benefits of nonspherical particles in terms of stability and resistance to nonspecific phagocytosis by macrophages as well as enhanced targeted binding [123]. Apart from the shape of the nanoparticles, an important aspect is the source and the method for applying the lipid coating to the nanoparticles. The lipids can be derived from the membranes of cells using top-down biomimetic procedures that first require disruption of the integrity of the cells to obtain cell membranes and then to fuse the membranes with the core nanoparticles by applying external forces such as extrusion or sonication. This type of coating is thought to inherently mimic the surface properties of the source cells for the benefits of superior biocompatibility, decreased uptake by macrophage cells, prolonged circulation penetration lifetimes, and enhanced tumor penetration. However, to achieve such benefits from such cell-membrane-derived lipids, the top-down biomimetic processing would need to maintain the normal biological nanostructure and orientation of the cell membranes, in addition to ensuring complete coverage of the nanoparticles. However, it has been demonstrated that up to 90% of nanoparticles are only partially coated using that type of top-down cell-membrane-derived processing [124].

An interesting alternative to producing cell-membrane-coated nanoparticles is to harvest and utilize the extracellular vesicles (EVs) that are naturally formed as compartments derived from cell membranes and released by cells. EVs are spherical membrane compartments, which contain lipids, proteins, and various nucleic acid species from their parent cells, acting as critical mediators between cells that regulate both physiological and pathological conditions in the body. Cells in the body constantly release EVs to deliver signaling molecules (such as proteins or DNA) to adjacent cells or in distant organs. There are three categories of EVs that are based on their biogenesis and sizes, which include exosomes, microvesicles, and apoptotic bodies. EVs have been utilized in diagnosis and drug delivery due to their exceptional biocompatibility, intrinsic targeting of specific molecules and maintenance of long circulation times. Despite the advantages of EVs, there a major barriers to be overcome before there can be EVs-based therapies or diagnostics. These barriers include (i) the difficulty to develop reproducible and scalable approaches to generate clinic-grade EVs, (ii) the appropriate environment (e.g., pH, temperature, buffered solutions) to avoid instability and aggregation of EVs especially when proteins are retained as cargo in the EVs, and (iii) the difficulty in targeting EVs to specific sites and to avoid unwanted clearance of the EVs from the body [125]. To overcome the barrier of poor scalability due to low yields, an approach has been to harvest giant membrane vesicles (GMVs, 5–22 μm in diameter) to be used as vectors to deliver multiple therapeutic drugs and carry out combinational phototherapy for targeted cancer treatment. The GMVs can be produced with high yields in a biocompatible buffer, thus making them compatible for a wide range of biological applications without separation [126].

The usefulness of lipid-coated nanoparticles for therapies has been demonstrated for diseases in the central nervous system (CNS), for cardiovascular disease, arthritis, and various types of cancers. The complexity of the central nervous system (CNS), its limited self-repairing capacity, and the ineffective delivery of most CNS drugs to the brain contribute to the irreversible and progressive nature of many neurological diseases and also the severity of the outcome. Therefore, neurological disorders belong to the group of pathologies with the greatest need of new technologies for diagnostics and therapeutics. Lipid-coated nanoparticles offer promising approaches for the treatment of neurological disorders. The sudden decrease of blood supply to the brain during a stroke causes a wide range of symptoms including numbness of the face, arms and legs, confusion, and aphasia. Stroke is one of the leading causes of death and disability worldwide. Ischemic stroke, or cerebral ischemia, is characterized by disruption of blood flow to the brain as a result of an obstruction in the arteries that nourish certain areas of the brain, leading to a cascade of physiopathological events in the CNS, including local blood brain barrier (BBB) disruption. However, to date, the only treatment approved by the FDA for this pathology is based on the potentially toxic tissue plasminogen activator. The techniques currently available for the diagnosis of stroke also lack sensitivity. Biomimetic nanoparticles can provide effective drug delivery to the brain and neuroprotection with the incorporation of targeting ligands on their surface for brain endothelial cells and ischemic tissues receptors. These biomimetic nanoparticles provide a way to cross the blood-brain-barrier and deliver molecules for therapy, biosensing, and imaging [127].

Cardiovascular disease is another major cause of death across the global population. Atherosclerosis, one of its most common drivers, is characterized by the gradual buildup of arterial plaque over time, which can ultimately lead to life-threatening conditions. Given the impact of the disease on public health, there is a great need for effective and noninvasive imaging modalities that can provide valuable information on its biological underpinnings during development. Biomimetic nanoparticles prepared with a coating of platelet membrane around a synthetic nanoparticulate core create a system that is capable of delivering therapeutic molecules to activated endothelium, foam cells, and collagen. The biomimetic nanoparticles can also target subclinical regions of arteries susceptible to plaque formation, which then provides a real-time detection system by incorporating into the biomimetic nanoparticles a magnetic resonance imaging contrast agent. Such a diagnostic strategy allows better assessment of the development of atherosclerosis, offering additional information to help clinicians better manage the disease [128].

Arthritis is the inflammation and tenderness of the joints because of various metabolic, infectious, or constitutional reasons. Different cell-membrane-coated biomimetic nanoparticle, EVs, and platelet-based systems provide opportunities for an emerging and efficient class of therapeutics to treat arthritis [129]. The high level of reactive oxygen species (ROS) in the rheumatoid arthritis (RA) microenvironment (RAM) and its persistent inflammatory nature can promote damage to joints, bones, and the synovium. Membrane vesicles produced from the cell membranes of macrophages (MMVs) have been used to deliver tacrolimus in a mouse model to suppress the progression of RA [130]. MMVs have also been demonstrated, in a mouse model of RA, to be very efficient in delivering small interfering RNAs (siRNAsT/I) and Prussian blue nanoparticles (PBNPs) to silence the expression of proinflammatory cytokines TNF-α/IL-6 and scavenge the ROS associated with RAM [131].

The effective treatment for cancers is a major incentive to drive the development of more efficient therapies. The usefulness of lipid-coated nanoparticles is in improving the targeting and efficiency of delivering cancer medications. An example is the development of bioengineered cell membrane nanovesicles (BCMNs) that improve the delivery of oncolytic adenoviruses to treat tumors [132]. Oncolytic virotherapy is a therapeutic approach to cancer treatment that utilizes viruses that selectively replicate in tumor cells. This has shown promising results in both preclinical studies and clinical trials. However, there are issues related to oncolytic virus delivery via the bloodstream, such as nonspecific sequestration, preexisting antivirus immunity, and the innate immune response, which pose major barriers to the development of oncolytic virotherapy. Lv et al. [132] developed an efficient strategy for fabricating a new class of bioengineered cell membrane nanovesicles (BCMNs) that can harbor different targeting ligands via natural biosynthetic procedures. The BCMNs may provide a clinical basis for improving oncolytic virotherapy by overcoming undesired antiviral immunity and enhancing cancer cell selectivity via biomimetic synthesis approaches [132].

Another type of cell membrane vesicle was formed by fusing a bacterial outer membrane vesicle (OMV) with a B16-F10 cancer cell (CC) membrane and then coating that membrane structures onto hollow polydopamine (HPDA) nanoparticles. This coated nanoparticle was produced to harness the advantage of OMV immunotherapy together with HPDA-mediated photothermal therapy (PTT) to improve the antitumor cell membrane efficacy toward melanoma. When injected intravenously via the tail vein, HPDA@[OMV-CC] NPs homogeneously target melanoma and activate the immune response by rapidly stimulating dendritic cell (DC) maturation in lymph nodes in the vaccinated mice. Results in mice showed that the antitumor immune response and PTT reciprocally potentiate the therapeutic ability and fully eradicate melanoma without notable adverse effects. The homogeneous-target and immune activation hybrid biomimetic membrane provides the adaptability to various synergistic therapeutic and imaging applications by incorporating payload with application-specific functions [133].

The incorporation of aptamers in the membranes of leucoyte-derived nanovesicles has provided a novel approach for the detection and capture of circulating tumor cells (CTCs) [134]. These aptamer-nanovesicles were incorporated into the microfluidic chip (FLASH), which created a biomimetic nanointerface that recapitulated the structure and function of a natural biomembrane, affording antiadsorption and soft interface to reduce blood cell adsorption and cell damage. In addition, the microfluidic platform provided a size-selective enhancement of cell-interface collision frequency. By combining the nanoscaled cell-interface interaction with the microscaled flow manipulation, that FLASH chip was capable of isolating CTCs from 17 cancer patient blood samples with high-efficiency, high-specificity, and high-viability. The approach of using aptamers to improve the capture of cells was also used in a system that was based on an extracted cell membrane of MCF-7 tumor cells attached to a graphene supporting membrane [135]. In that system, multivalent aptamers were conjugated to the surface of the extracted MCF-7 cell membrane to provide specific binding sites for targeting epithelial cell adhesion molecule (EpCAM), which is overexpressed in most solid cancers and has been demonstrated previously as a target for the isolation of CTCs [136]. That system was capable to efficiently capturing MCF-7 tumor cells spiked in blood samples but was not tested for the capture of native CTCs.

The medications used to treat cancers often suffer from the phenomenon of multi-drug resistance (MDR), which remains a dominant cause in cancer chemotherapy failure. The major factor currently considered to be responsible for MDR is the upregulation of P-glycoprotein expression in cancer cells. The current generation of P-glycoprotein inhibiting drugs are highly toxic and not approved for clinical use. The discovery of new-generation drugs to inhibit P-glycoprotein requires a reliable testing platform, and for that novel drug-screening system has been developed [137]. This system comprises inside-out oriented extracellular vesicles biomimetic nanomaterials, so that the orientation of incorporated P-glycoprotein can be controlled to ensure efficient screening of potential P-glycoprotein inhibitors derived from natural products. Zhang et al. [137] claim that this system could also facilitate the screening of inhibitors for other complex membrane proteins, compared to the existing alternate approaches of bidirectional transcellular transport assays and the absence of any high resolution protein crystal structures that could be used for in silico virtual screening assays [137]. That type of approach to develop novel biomembrane-based screening assays has been applied to ligand-specific cellular entry systems relevant to head and neck cancer [138]. Navarro-Palomares et al. [138] developed a biomimetic nanosystem that comprised membrane-coated nanoparticles with a recombinant chimera protein containing the innocuous B domain of the Shiga toxin genetically fused to a nanomaterial-binding sequence. This globoside is typically found in immature endothelial cells of the intestine and is also present on the surfaces of many cancerous/precancerous cells of the digestive tract. The chimera mimicked the molecular cues found in the Shiga toxin to target nanomaterials to head and neck cancer cells bearing the globotriaosylceramide receptor (known as GB3 or CD77). The results from Navarro-Palmares et al demonstrate that nanomaterials functionalized with this ligand recognize GB3 +ve cells reproducing the cellular responses activated by the full-length toxin to sneak into the cellular cytoplasm [138]. This is another biomimetic nanostructured system that demonstrates how custom biofunctionalization methods can transform inert nanomaterials into hybrid nanosystems capable of identifying specific receptors predictably triggering specific cellular responses as dictated by the proteins in their membrane coating.

An "Ideal" Lipid Bilayer Membrane to Enhance the Performance of Implanted Medical Devices

Enhanced performance of an implanted medical device requires an interface that does not induce a foreign-body reaction. A lipid bilayer membrane added to the surface of an implanted medical device could provide such interface that does not induce an inflammatory and immune response. In considering whether a tBLM can provide a robust "stealth" interface between an implanted medical device and the body, we need to consider what are the major factors that trigger an immune response and an inflammatory response to remove foreign particle (or molecules). This is the self-nonself recognition that we discussed in the previous sections [82].

We can explore this idea of utilizing a lipid bilayer added to medical devices by returning to the hypothesis stated in earlier sections. The biological rationale for this hypothesis is to minimize the exposure of the body to components of a biomaterial that can stimulate a foreign-body reaction, as we summarized in Fig. 4.4 [56]. Further, the hypothesis addresses the need to minimize the excessive complement activation that could contribute to tissue damage, in an analogous way to the challenge of trying to inhibit the excessive complement activation that occurs in chronic, low-grade, and inflammatory processes of aging [139]. The hypothesis is that a lipid bilayer membrane could function as a "stealth" membrane interface to render an implanted medical device as "self" to minimize excessive complement activation for the immune system by incorporating at least one membrane-bound protein of CD35, CD46, CD55, or CD59. Additionally, the lipid bilayer membrane would be further coated by a layer of molecules to mimic the natural glycocalyx.

We can consider the merits of this hypothesis by examining the literature to consider the design of the lipids that could comprise an "ideal" lipid bilayer membrane. As discussed in earlier sections, such an "ideal" lipid bilayer membrane would need to be tethered securely to the surface of the implanted medical device. Moreover, the choice of lipids would be influenced by the known binding characteristics of complement factors to function as lipid opsonins. For example, adiponectin C1q family protein and other C1q/TNF-related proteins (CTRPs) bind anionic phospholipids and sphingolipids, including phosphatidic acid, phosphatidylserine (PS), phosphatidylinositol phosphate (PIP), phosphatidylinositol 4,5-bisphosphate (PIP_2), phosphatidylinositol (3,4,5)-trisphosphate (PIP_3), cardiolipin (CL), sulfatide (ST), ceramide-1-phosphate (Cer1P), and dihydroceramide-1-phosphate (DHCer1P) [140].

It has been known for some time that phospholipids need to be anchored on the surface of an implanted device in order to provide a stable interface [79]. Indeed, a strategy was published to design a chemically anchored phospholipid monolayer that could be used as coating materials for biomedical implants. This strategy included the preparation of a mono-acrylated phospholipid (1-palmitoyl-2-[12-(acryloyloxy)-dodecanoyl]-sn-glycero-3-phosphocholine; acryloyl-PC) and acrylated polymer (poly(octadecylacrylate-co-4-acryloyloxy butylacrylate)), which was synthesized by the acrylation of poly(octadecyl acrylate-co-hydroxybutyl acrylate, poly(OA-co-HA)) with acryloyl chloride. The chemically anchored phospholipid monolayer was prepared by using in situ photopolymerization of a pre-assembled phospholipid monolayer, produced by lipid vesicle fusion, onto the acrylated polymer coated silicon wafer [141]. Kim et al. [141] demonstrated that their chemically anchored phospholipid monolayer was stable for 2 weeks in aqueous conditions, compared to a physically adsorbed phospholipid monolayer as a control that was removed within one day by aqueous conditions.

A monolayer of 2-oleoyl-1-palmitoyl-sn-glycero-3-phosphoethanolamin (POPE) was deposited to titanium plates using Langmuir-Blodgett film deposition and implanted intramuscularly in rats for up to 56 days to assess the local inflammatory reactions to a POPE-based biomimetic surface. The POPE monolayers were either (i) attached covalently using octadecylphosphonic acid (OPA) or (ii) directly to the titanium plate, and the local inflammatory response was compared to titanium (uncoated) or to titanium coated only with OPA surface. There were no significant differences among all four titanium plates, with the only conclusion that the concept of coating an implant material using a biomembrane mimic based on the phospholipid POPE does not have a negative influence on the short- and long-term inflammatory reactions in the peri-implant tissue [142]. POPE has been shown to be an anionic phospholipid at pH 7.4, and not neutral [143].

An interesting approach to synthesize a stable phospholipid membrane has been to modify polymers with phospholipids to mimic the surface of a cell membrane. In the 1970s, the concept was proposed to synthesize a polymer/phospholipid structure using 2-methacryloyloxyethyl phosphorylcholine (MPC), which is a methacrylate monomer with a phosphorylcholine (PC) group. However, the practical realization of MPC was only achieved by the pioneering work of Ishihara, Ueda, and Nakabayashi in 1990, who reported a method to synthesize a methacrylate monomer having the phospholipid polar group, 2-methacryloyloxyethyl phosphorylcholine (MPC) by copolymerization with n-butyl methacrylate (BMA). Polymer membranes were prepared from the poly(MPC-*co*-BMA) by a solution casting method [144]. These MPC polymers have been used for many biointerfacial applications, predominately to create antifouling surfaces, and platforms of biomolecules and living cells [145]. More recently poly2-Methacryloyloxyethyl phosphorylcholine (pMPC) has been combined with two other zwitterionic polymers, polysulfobetaine methacrylate (pSBMA), and polycarboxybetaine methacrylate (pCBMA) to form self-assembled monolayers on blood-bearing medical devices. Although these more recent coatings have been highly successful in significantly reducing levels of proteins adsorption and platelet adhesion during in vitro studies on the order of hours and at reducing thrombus formation for up to a few days in implanted medical devices [146].

With any of these modified polymer/phospholipid materials, the remaining major challenge is the long-term performance. The difficulty with the modified polymer approach is the surface roughness due to the polymer and the inability to properly coat the phospholipids with an additional protective membrane of a glycocalyx mimic. Nonetheless, a recent approach to utilize a combination of block co-polymerization and layer-by-layer techniques has fabricated a multifunctional coating using a combination of phospholipids-based polymers, type I collagen (Col-I) and Arg-Glu-Asp-Val (REDV) peptide. Furthermore, the results of variety of cell cultures indicated that this phospholipid/Col-I/REDV coating could not only inhibit platelet adhesion and activation but also inhibit SMCs proliferation and promote ECs proliferation in short-term cell culture experiments [147]. This membrane did not have a bilayer structure nor thickness and so would not provide a suitable approach for incorporating membrane proteins. This would not provide the type of membrane to mimic a functional cell membrane and hence would have difficulty for long-term usage to coat implanted medical device to provide an appropriate "self" recognition interface for long-term implantation.

A structured lipid bilayer was added to a nanostructured coating of poly(tannic acid) used to bind a nanostructured layer-by-layer integration black phosphorous as a photothermal layer. The lipid bilayer was derived directly from red blood cells. The poly(tannic acid) layer immobilized and stabilized the subsequent black phosphorous and cell membrane layers on implants. The top cell membrane layer served as a modulator for precise local energy management, additionally acting as an extra protector for BP nanosheets. Cell culture tests showed that the lipid bilayer membrane provided an excellent biocompatible interface and contributed to the enhanced stability of the black phosphorous nanosheets against oxidation and degradation that reduced ROS generation within cells. Consequently, these assays underscored the favorable cytocompatibility of the lipid bilayer membrane coating and confirmed its value to provide a biocompatible candidate for in vivo applications [148].

In concluding this chapter, we can keep in mind that the MPC-modified polymer approach provides a promising concept for assembling stable mimics of lipid bilayers. Nonetheless, even if MPC produces a stable polymer/phospholipid material, it does not provide a "self" recognition for long-term implanted applications. Perhaps a better utilization for future designs of implanted devices is to create 3D matrixes of MPC to provide scaffolds for cellular growth to provide artificial organ applications. The "self-nonself" discrimination requires interfaces that can modulate the binding and activation of complement factors, leading to the minimization of inflammatory responses that occur, for example, particularly chronic low-level responses. The "self-nonself" discrimination relies upon a functional cell-membrane, which needs to be mimicked in an interface for implanted medical devices. Even if a self-assembled membrane of MPC does produce a thin membrane to modulate cell attachment, it is not clear that the MPC approach can produce a mimic of the cell membrane bilayer that is both a "tight" barrier and capable of incorporating the required membrane proteins to mimic the "self" recognition that is necessary to protect and enable long-term function of implanted medical devices.

We can conclude that a better approach to create an interface is to provide "self" recognition by using a stable tBLM with incorporated membrane proteins to mimic the biological glycocalyx that provides the natural protection for the native biological cell membrane. We reiterate an hypothesis for a concept to minimize the chronic complement activation and inflammatory foreign-body reactions to implanted medical devices. That hypothesis is constructed to produce an "ideal" lipid bilayer membrane to enhance the performance of implanted medical devices. The hypothesis is that a lipid bilayer membrane could function as a "stealth" membrane interface to render an implanted medical device as "self" to minimize excessive complement activation for the immune system by incorporating at least one membrane-bound protein of CD35, CD46, CD55, or CD59. Additionally, the lipid bilayer membrane would be further coated by a layer of molecules to mimic the natural glycocalyx. This concept and hypothesis is necessary to ensure that implanted medical devices have a functional activity in the body. In Chap. 2, we discussed this need for next-generation implanted medical devices in the context of "symbio-bots." To ensure long-term functional activity means that the interface between a "symbio-bot" and the body needs to remain biologically functional. The important aspect of the interface is for it to be recognized as "self," which would provide the situation for the interface of the implanted medical device to maintain biological functions, such as ion and molecule transport. That is, a next-generation "symbio-bot" implanted medical device will provide the long-term function to be able to replace damaged organs and tissues in the body.

References

1. Bricarello, D. A., Smilowitz, J. T., Zivkovic, A. M., German, J. G., & Parikh, A. N. (2011). Reconstituted lipoprotein: A versatile class of biologically-inspired nanostructures. *ACS Nano, 5*, 42–57.
2. Wendell, D. W., Patti, J., & Montemagno, C. D. (2006). Using biological inspiration to engineer functional nanostructured materials. *Small, 2*, 1324–1329.
3. Richard, P., Pitard, B., & Rigaud, J. L. (1995). ATP synthesis by the F0F1-ATPase from the thermophilic bacillus PS3 co-reconstituted with bacteriorhodopsin into liposomes. *Journal of Biological Chemistry, 270*, 21571–21578.
4. Pitard, P., Richard, M., DuÇarch, G., & Girault, J. R. (1996). ATP synthesis by the F,F, ATP synthase from thermophilic Bacillus PS3 reconstituted into liposomes with bacteriorhodopsin. *European Journal of Biochemistry, 235*, 769–778.
5. Gong, Y. K., & Winnik, F. M. (2012). Strategies in biomimetic surface engineering of nanoparticles for biomedical applications. *Nanoscale, 4*, 360–368.
6. Alcaraz, J. P., Cinquin, P., & Martin, D. K. (2018). Tackling the oncept of the symbiotic implantable medical devices with nanotechnologies. *Biotechnology Journal, 13*, 1800102.
7. Liu, Y., Kim, E., Ghodssi, R., Rubloff, G. W., Culver, J. N., Bentley, W. E., & Payne, G. F. (2010). Biofabrication to build the biology-device interface. *Biofabrication, 2*, 022002.
8. Goes, A., & Fuhrmann, G. (2018). Biogenic and biomimetic carriers as versatile transporters to treat Infections. *ACS Infectious Diseases, 4*, 881–892.

9. Guimarães, C. F., Soto, F., Wang, J., Akin, D., Reis, R. L., & Demirci, U. (2022). Engineered living bioassemblies for biomedical and functional material applications. *Current Opinion in Biotechnology, 77*, 102756.
10. Govindasamy, N., Long, H., Jeong, H. W., Raman, R., Özcifci, B., Probst, S., Arnold, S. J., Riehemann, K., Ranga, A., Adams, R. H., Trappmann, B., & Bedzhov, I. (2021). 3D biomimetic platform reveals the first interactions of the embryo and the maternal blood vessels. *Developmental Cell, 56*, 3276–3287.
11. Fang, R. H., Jiang, Y., Fang, J. C., & Zhang, L. (2017). Cell membrane-derived nanomaterials for biomedical applications. *Biomaterials, 128*, 69–83.
12. Mindick, M., & Patwelt, H. I. (1960). Production of ion selective permeable membrane and composition therefor. *US, 2*(957), 206.
13. Decher, G., & Hong, J. D. (1991). Buildup of ultrathin multilayer films by a self-assembly process, 1 consecutive adsorption of anionic and cationic bipolar amphiphiles on charged surfaces. *Makromolekulare Chemie-Macromolecular Symposia, 46*, 321–327.
14. Decher, G., Hong, J. D., & Schmitt, J. (1992). Buildup of ultrathin multilayer films by a self-assembly process. III. Consecutively alternating adsorption of anionic and cationic polyelectrolytes on charged surfaces. *Thin Solid Films, 210-211*, 831–835.
15. Kügler, R., & Knoll, W. (2002). Polyelectrolyte-supported lipid membranes. *Bioelectrochemistry, 56*, 175–178.
16. Chen, J., Köhler, R., Gutberlet, T., Möhwald, H. R., & Krastev, R. (2009). Asymmetric lipid bilayer sandwiched in polyelectrolyte multilayer films through layer-by-layer assembly. *Soft Matter, 5*, 228–233.
17. Stidder, B., Alcaraz, J. P., Liguori, L., Khalef, N., Bakri, A., Watkins, E., Cinquin, P., & Martin, D. K. (2012). Biomimetic membrane system composed of a composite interpenetrating hydrogel film and lipid bilayer. *Advanced Functional Materials, 20*, 4259–4267.
18. Nolte, M., & Fery, A. (2006). Freestanding polyelectrolyte multilayers as functional and construction elements. *IEE Proceedings Nanobiotechnology, 153*, 112–120.
19. Donath, E., Sukhorukov, G. B., Caruso, F., Davis, S. A., & Möhwald, H. (1998). Novel hollow polymer shells by colloid-templated assembly of polyelectrolytes. *Angewandte Chemie International Edition, 37*, 2201.
20. Battle, A. R., Valenzuela, S. M., Mechler, A., Nichols, R. J., Praporski, S., di Maio, I. L., Islam, H., Girard-Egrot, A. P., Cornell, B. A., Prashar, J., Caruso, F., Martin, L. L., & Martin, D. K. (2009). Novel engineered ion channel provides controllable ion permeability for polyelectrolyte microcapsules coated with a lipid membrane. *Advanced Functional Materials, 19*, 201–208.
21. Blodgett, K. B. (1935). Films built by depositing successive monomolecular layers on a solid surface. *Journal of the American Chemical Society, 57*, 1007–1022.
22. Blodgett, K. B., & Langmuir, I. (1937). Built-up films of barium stearate and their optical properties. *Physical Review, 51*, 964–982.
23. Blodgett, K. B. (1939). A method of extinguishing the reflection of light from glass. *Science, 89*, 60–61.
24. Johonnott, E. S. (1899). Thickness of the black spot in liquid films. *Philosophical Magazine Journal of Science, Series, 5*(47), 501–522.
25. Johonnott, E. S. (1906). The black spot in thin liquid films. *Philosophical Magazine Journal of Science, Series, 6*(11), 746–753.
26. Folch, J., & Lees, M. (1951). Proteolipides, a new type of tissue lipoproteins—Their isolation from brain. *Journal of Biological Chemistry, 191*, 807–817.
27. Mueller, P., Rudin, D. O., Ti Tien, H., & Wescott, W. C. (1962). Reconstitution of cell membrane structure in vitro and its transformation into an excitable system. *Nature, 194*, 979–980.
28. Mueller, P., Rudin, D. O., Ti Tien, H., & Wescott, W. C. (1962). Reconstitution of cell membrane structure in vitro. *Circulation, 26*, 1167–1171.
29. Bangham, A. D., Standish, M. M., & Watkins, J. C. (1965). Diffusion of univalent ions across the lamellae of swollen phospholipids. *Journal of Molecular Biology, 13*, 238–252.

30. Ti Tien, H., & Diana, A. L. (1968). Biomolecular lipid membranes: A review and a summary of some recent studies. *Chemistry and Physics of Lipids, 2*, 55–101.
31. Ti Tien, H. (1988). Bilayer lipid membrane-based electrochemical biosensors. *Journal of Clinical Laboratory Analysis, 2*, 256–264.
32. Ottova-Leitmannova, A., & Ti Tien, H. (1992). Bilayer lipid membranes: An experimental system for biomolecular electronic devices development. *Progress in Surface Science, 41*, 337–445.
33. Ottova, A., Tvarozek, V., Racek, J., Sabo, J., Ziegler, W., Hianik, T., & Tien, H. T. (1997). Self-assembled BLMs: Biomembrane models and biosensor applications. *Supramolecular Science, 4*, 101–112.
34. Feller, S. E. (2000). Molecular dynamics simulations of lipid bilayers. *Current Opinion in Colloid & Interface Science, 5*, 217–223.
35. Castellana, E. T., & Cremer, P. S. (2006). Solid supported lipid bilayers: From biophysical studies to sensor design. *Surface Science Reports, 61*, 429–444.
36. Martin, D. K. (Ed.). (2007). *Nanobiotechnology of Biomimetic Membranes*. Springer.
37. Köper, I. (2007). Insulating tethered bilayer lipid membranes to study membrane proteins. *Molecular BioSystems, 3*, 651–657.
38. Mouritsen, O. G. (2011). Lipids, curvature, and nano-medicine. *European Journal of Lipid Science and Technology, 113*, 1174–1187.
39. Maccarini, M., Watkins, E. B., Stidder, B., Alcaraz, J. P., Cornell, B. A., & Martin, D. K. (2016). Nanostructural determination of a lipid bilayer tethered to a gold substrate. *European Physical Journal E, 39*, 123.
40. Köhler, S., Fragneto, G., Alcaraz, J. P., Nelson, A., Martin, D. K., & Maccarini, M. (2021). Nanostructural characterization of cardiolipin-containing tethered lipid bilayers adsorbed on gold and silicon substrates for protein incorporation. *Langmuir, 37*, 8908–8923.
41. Fragneto, G., Delhom, R., Joly, L., & Scoppola, E. (2018). Neutrons and model membranes: moving towards complexity. *Current Opinion in Colloid & Interface Science, 38*, 108–120.
42. Kim, T., Lee, K. I., Morris, P., Pastor, R. W., Andersen, O. S., & Im, W. (2012). Influence of hydrophobic mismatch on structures and dynamics of gramicidin A and lipid bilayers. *Biophysical Journal, 102*, 1551–1560.
43. Maccarini, M., Gayet, L., Alcaraz, J. P., Liguoria, L., Stidder, B., Cortès, S., Watkins, E., Lenormand, J. L., & Martin, D. K. (2017). Function of recombinant OprF from Pseudomonas aeruginosa stably incorporated in supported lipid bilayers. *Langmuir, 33*, 9988–9996.
44. Spinozzi, F., Alcaraz, J. P., Ortore, M. G., Gayet, L., Radulescu, A., Martin, D. K., & Maccarini, M. (2022). Small-angle neutron scattering reveals the nanostructure of liposomes with embedded OprF porins of Pseudomonas aeruginosa. *Langmuir, 38*, 15026–15037.
45. Soranzo, T., Martin, D. K., Lenormand, J. L., & Watkins, E. B. (2017). Coupling neutron reflectivity with cell-free protein synthesis to probe membrane protein structure in supported bilayers. *Scientific Reports, 7*, 3399.
46. Langer, R., & Folkman, J. (1976). Polymers for the sustained release of proteins and other macromolecules. *Nature, 263*, 797–800.
47. Langer, R., Brem, H., & Tapper, D. (1981). Biocompatibility of polymeric delivery systems for macromolecules. *Journal of Biomedical Materials Research, 15*, 267–277.
48. Ratner, B. D., Horbett, T., Hoffman, A. S., & Hauschka, S. D. (1975). Cell adhesion to polymeric materials: Implications with respect to biocompatibility. *Journal of Biomedical Materials Research, 9*, 407–422.
49. Ratner, B. D., Weathersby, P. K., Hoffman, A. S., Kelly, M. A., & Scharpen, L. H. (1978). Radiation-grafted hydrogels for biomaterial applications as studied by the ESCA technique. *Journal of Applied Polymer Science, 22*, 643–664.
50. Williams, D. F. (1976). Biomaterials and biocompatibility. *Medical Progress through Technology, 4*, 31–42.

51. Anderson, J. M., Hiltner, A., Schodt, K., & Woods, R. (1972). Biopolymers as biomaterials: mechanical properties of gamma-benzyl-L-glutamate-L-leucine copolymers. *Journal of Biomedical Materials Research, 6*, 25–35.
52. Anderson, J. M., Gibbons, D. F., Martin, R. L., Hiltner, A., & Woods, R. (1974). The potential for poly-alpha-amino acids as biomaterials. *Journal of Biomedical Materials Research, 8*, 197–207.
53. Horne, R. A. (1968). *N_. TR-34, Little A*. D INC Cambridge, MA, p30.
54. Doshi, G. R., Sreekumaran, C., Mulay, C. D., & Patel, B. (1969). Ashing procedures for biomaterial. *Current Science, 38*, 206–208.
55. Williams, D. F. (2009). On the nature of biomaterials. *Biomaterials, 30*, 5897–5909.
56. Williams, D. F. (2014). There is no such thing as a biocompatible material. *Biomaterials, 35*, 10009–10014.
57. Williams, D. F. (2016). Biocompatibility pathways: Biomaterials-induced sterile inflammation, mechanotransduction, and principles of biocompatibility control. *ACS Biomaterials Science and Engineering, 3*, 2–35.
58. Janeway, C. A., Travers, P., Walport, M., & Shlomchik, M. J. (2001). *Immunobiology, 5th edition: The immune system in health and disease*. Garland Science Publishing. ISBN-10:0-8153-3642-X.
59. van Loon, S. L. M., Smits, A. I. P. M., Baaijens, F. P. T., & Bouten, C. V. C. (2013). The immune response in In Situ tissue engineering of aortic heart valves. In E. Aikawa (Ed.), *Calcific aortic valve disease* (pp. 207–245). InTech. https://doi.org/10.5772/54354
60. Bryers, J. D., Giachelli, C. M., & Ratner, B. D. (2012). Engineering biomaterials to integrate and heal: the biocompatibility paradigm shifts. *Biotechnology and Bioengineering, 109*, 1898–1911.
61. Anderson, J. M. (2011). Biological responses to materials. *Annual Review of Materials Research, 31*, 81–110.
62. Zebda, A., Alcaraz, J. P., Vadgama, P., Shleev, S., Minteer, S. D., Boucher, F., & Martin, D. K. (2018). Challenges for successful implantation of biofuel cells. *Bioelectrochemistry, 2*, 57–72.
63. Ratner, B. D., Hoffman, A. S., Schoen, F. J., & Lemons, J. E. (2004). *Biomaterials science: An introduction to materials in medicine* (pp. 296–3014). Academic. ISBN: 0-12-582460-2.
64. Voskerician, G., Shive, M. S., Shawgo, R. S., Von Recum, H., Anderson, J. M., Cima, M. J., & Langer, R. (2003). Biocompatibility and biofouling of MEMS drug delivery devices. *Biomaterials, 24*, 1959–1967.
65. Jon, S., Seong, J., Khademhosseini, A., Tran, T. N. T., Laibinis, P. E., & Langer, R. (2003). Construction of nonbiofouling surfaces by polymeric self-assembled monolayers. *Langmuir, 19*, 9989–9993.
66. Brown, B. N., Londono, R., Tottey, S., Zhang, S., Kukla, K. A., Wolf, M. T., Daly, K. A., Reing, J. E., & Badylak, S. F. (2012). Macrophage phenotype as a predictor of constructive remodeling following the implantation of biologically derived surgical mesh materials. *Acta Biomaterialia, 8*, 978–987.
67. Sussman, E. M., Halpin, M. C., Muster, J., Moon, R. T., & Ratner, B. D. (2014). Porous implants modulate healing and induce shifts in local macrophage polarization in the foreign body reaction. *Annals of Biomedical Engineering, 42*, 1508–1516.
68. Anderson, J. M., Rodriguez, A., & Chang, D. T. (2008). Foreign body reaction to biomaterials. *Seminars in Immunology, 20*, 86–100.
69. Kou, P. M., & Babensee, J. E. (n.d.). Macrophage and dendritic cell phenotypic diversity in the context of biomaterials. *Journal of Biomedical Materials Research A, 96*, 239–260.
70. Helming, L., Tomasello, E., Kyriakides, T. R., Martinez, F. O., Takai, T., Gordon, S., & Vivier, E. (2008). Essential role of DAP12 signaling in macrophage programming into a fusion-competent state. *Science Signaling, 1*, ra11.

71. Ratner, B. D. (2001). A perspective on titanium biocompatibility. In D. M. Brunette, O. Tengvall, M. Textor, & P. Thomsen (Eds.), *Titanium in Medicine* (pp. 1–12). Springer. ISBN 3-540-66936-1.
72. Brunette, D. M. (1988). The effects of implant surface topography on the behavior of cells. *International Journal of Oral & Maxillofacial Implants, 3*, 231–246.
73. Ratner, B. D. (2016). A pore way to heal and regenerate: 21st century thinking on biocompatibility. *Regenerative Biomaterials, 3*, 107–110.
74. Puskas, J. E., & Luebbers, M. T. (2012). Breast implants: The good, the bad and the ugly. Can nanotechnology improve implants? *Wiley Interdisciplinary Reviews Nanomedicine and Nanobiotechnology, 4*, 153–168.
75. Singh, C., Wong, C. S., & Wang, X. (2015). Medical textiles as vascular implants and their success to mimic natural arteries. *Journal of Functional Biomaterials, 6*, 500–525.
76. Khalifehzadeh, R., Ciridon, W., & Ratner, B. D. (2018). Surface fluorination of polylactide as a path to improve platelet associated hemocompatibility. *Acta Biomaterialia, 78*, 23–35.
77. Li, S., Ju, Y., Zhou, J., Faria, M., Ang, C. S., Mitchell, A. J., Zhong, Q. Z., Zheng, T., Kent, S. J., & Caruso, F. (2022). Protein precoating modulates biomolecular coronas and nanocapsule-immune cell interactions in human blood. *Journal of Materials Chemistry B, 10*, 7607–7621.
78. Joh, D. Y., Zimmers, Z., Avlani, M., Heggestad, J. T., Aydin, H. B., Ganson, N., Kumar, S., Fontes, C. M., Achar, R. K., Hershfield, M. S., Hucknall, A. M., & Chilkoti, A. (2019). Architectural modification of monformal PEG-bottlebrush coatings minimizes anti-PEG antigenicity while preserving stealth properties. *Advanced Healthcare Materials, 8*(8), e1801177.
79. Kim, K., Kim, C., & Byun, Y. (2004). Biostability and biocompatibility of a surface-grafted phospholipid monolayer on a solid substrate. *Biomaterials, 25*, 33–41.
80. Willumeit, R., Schuster, A., Iliev, P., Linser, S., & Feyerabend, F. (2007). Phospholipids as implant coatings. *Journal of Materials Science Materials in Medicine, 18*(2), 367–380.
81. Santin, M., Rhys-Williams, W., O'Reilly, J., Davies, M. C., Shakesheff, K., Love, W. G., Lloyd, A. W., & Denyer, S. P. (2006). Calcium-binding phospholipids as a coating material for implant osteointegration. *Journal of the Royal Society Interface, 3*, 277–281.
82. Meri, S. (2016). Self-nonself discrimination by the complement system. *FEBS Letters, 590*, 2418–2434.
83. Voisin, T. B., Couves, E. C., Tate, E. W., & Bubeck, D. (2023). Dynamics and molecular interactions of GPI-anchored CD59. *Toxins, 15*, 430.
84. Teoh, C. W., Riedl Khursigara, M., Ortiz-Sandoval, C. G., Park, J. W., Li, J., Bohorquez-Hernandez, A., Bruno, V., Bowen, E. E., Freeman, S. A., Robinson, L. A., & Licht, C. (2023). The loss of glycocalyx integrity impairs complement factor H binding and contributes to cyclosporine-induced endothelial cell injury. *Frontiers in Medicine, 10*, 891513.
85. Kang, S., Onishi, S., Ling, Z., Inoue, H., Zhang, Y., Chang, H., Zhao, H., Wang, T., Okuzaki, D., Matsuura, H., Takamatsu, H., Oda, J., & Kishimoto, T. (2023). Gp130-HIF1α axis-induced vascular damage is prevented by the short-term inhibition of IL-6 receptor signaling. *PNAS, 121*, e2315898120.
86. Ye, J. J., Bian, X., Lim, J., & Medzhitov, R. (2020). Adipnectin and related C1q/TNF-related proteins bind selectively to anionic phospholipids and sphingolipids. *PNAS, 117*, 17381–17388.
87. Tan, L. A., Yu, B., Sim, F. C. J., Kishore, U., & Sim, R. B. (2010). Complement activation by phospholipids: the interplay of factor H and C1q. *Protein & Cell, 1*, 1033–1049.
88. Cornell, B. A., Braach-Maksvytis, V. L., King, L. G., Osman, P. D., Raguse, B., Wieczorek, L., & Pace, R. J. (1997). A biosensor that uses ion channel switches. *Nature, 387*, 580–583.
89. Andersson, J., & Köper, I. (2016). Tethered and polymer supported bilayer lipid membranes: structure and function. *Membranes, 6*, 30.
90. Dupuy, A. D., & Engelman, D. M. (2008). Protein area occupancy at the center of the red blood cell membrane. *PNAS, 105*, 2848–2852.

91. Vockenroth, I. K., Ohm, C., Robertson, J. W. F., McGillivray, D. J., Lösche, M., & Köper, I. (2008). Stable insulating tethered bilayer lipid membranes. *Biointerphases, 3*, FA68–FA73.
92. Prashar, J., Sharp, P., Scarffe, M., & Cornell, B. (2007). Making lipid membranes even tougher. *Journal of Materials Research, 22*, 2189–2194.
93. Cranfield, C. G., Cornell, B. A., Grage, S. L., Duckworth, P., Carne, S., Ulrich, A. S., & Martinac, B. (2014). Transient potential gradients and impedance measures of tethered bilayer lipid membranes: Pore forming peptide insertion and the effect of electroporation. *Biophysical Journal, 106*, 182–189.
94. Plant, A., Gueguetchkeri, M., & Yap, W. (1994). Supported phospholipid/alkanethiol biomimetic membranes: insulating properties. *Biophysical Journal, 67*, 1126–1133.
95. Raguse, B., Braach-Maksvytis, V., Cornell, B. A., King, L. G., Osman, P. D. J., Pace, R. J., & Wieczorek, L. (1998). Tethered lipid bilayer membranes: Formation and ionic reservoir characterization. *Langmuir, 14*, 648–659.
96. Krishna, G., Schulte, J., Cornell, B. A., Pace, R. J., Wieczorek, L., & Osman, P. D. (2001). Tethered bilayer membranes containing ionic reservoirs: The interfacial capacitance. *Langmuir, 17*, 4858–4866.
97. Krishna, G., Schulte, J., Cornell, B. A., Pace, R. J., & Osman, P. D. (2003). Tethered bilayer membranes containing ionic reservoirs: Selectivity and conductance. *Langmuir, 19*, 2294–2305.
98. Naumann, R. L., Schiller, S. M., Giess, F., Grohe, B., Hartman, K. B., Kärcher, I., Köper, I., Lübben, J., Vasilev, K., & Knoll, W. (2003). Tethered lipid bilayers on ultraflat gold surfaces. *Langmuir, 19*, 5435–5443.
99. Schiller, S. M., Naumann, R., Lovejoy, K., Kunz, H., & Knoll, W. (2003). Archaea analogue thiolipids for tethered bilayer lipid membranes on ultrasmooth gold surfaces. *Angewandte Chemie International Edition, 42*, 208–211.
100. Andersson, J., Knobloch, J. J., Perkins, M. V., Holt, S. A., & Köper, I. (2017). Synthesis and characterization of novel anchor lipids for tethered bilayer lipid membranes. *Langmuir, 33*, 4444–4451.
101. McGillivray, D. J., Valincius, G., Vanderah, D. J., Wilma, F., Woodward, J. T., Heinrich, F., Kasianowicz, J. J., & Loesche, M. (2007). Molecular-scale structural and functional characterization of sparsely tethered bilayer lipid membranes. *Biointerphases, 2*, 21–33.
102. Heinrich, F., Ng, T., Vanderah, D. J., Shekhar, P., Mihailescu, M., Nanda, H., & Loesche, M. (2009). A new lipid anchor for sparsely tethered bilayer lipid membranes. *Langmuir, 25*, 4219–4229.
103. Liefeith, K., Frant, M., Müller, U., Stenstad, P., Johnsen, H., & Schmid, R. (2018). Archaeal tetraether lipid coatings: A strategy for the development of membrane analog spacer systems for the site-specific functionalization of medical surfaces. *Biointerphases, 13*, 011004.
104. Squillace, O., Esnault, C., Pilard, J. F., & Brotons, G. (2019). Electrodes for membrane surface science: Bilayer lipid membranes tethered by commercial surfactants on electrochemical sensors. *ACS Sensors, 4*, 1337–1345.
105. Andersson, J., Fuller, M. A., Wood, K., Holt, S. A., & Köper, I. (2018). A tethered bilayer lipid membrane that mimics microbial membranes. *Physical Chemistry Chemical Physics, 20*, 12958–12969.
106. Lee, H. H., Gavutis, M., Ruzele, R., & Liedberg, B. (2018). Mixed self-assembled monolayers with terminal deuterated anchors: Characterization and probing of model lipid membrane formation. *Journal of Physical Chemistry B, 122*, 8201–8210.
107. Naumann, R., Baumgart, T., Graber, P., Jonczyk, A., Offenhausser, A., & Knoll, W. (2002). Proton transport through a peptidetethered bilayer lipid membrane by the H+-ATP synthase from chloroplasts measured by impedance spectroscopy. *Biosensors Bioelectronics, 17*, 25–34.
108. Chadli, M., Rebaud, S., Maniti, O., Tillier, B., Cortès, S., & Girard-Egrot, A. P. (2017). New tethered phospholipid bilayers Integrating functional G-protein-coupled receptor membrane proteins. *Langmuir, 33*, 10385–10401.

109. Rossi, C., Homand, J., Bauche, C., Hamdi, H., Ladant, D., & Chopineau, J. (2003). Differential mechanisms for calcium-dependent protein/membrane association as evidenced from SPR-binding studies on supported biomimetic membranes. *Biochemistry, 42*, 15273–15283.
110. Daniel, C., Sohn, K. E., Mates, T. E., Kramer, E. J., Rädler, J. O., Sackmann, E., Nickel, B., & Andruzzi, L. (2007). Structural characterization of an elevated lipid bilayer obtained by stepwise functionalization of a self-assembled alkenyl silane film. *Biointerphases, 2*, 109–118.
111. Hertrich, S., Stetter, F., Rühm, A., Hugel, T., & Nickel, B. (2014). Highly hydrated deformable polyethylene glycol-tethered lipid bilayers. *Langmuir, 30*, 9442–9447.
112. Atanasov, V., Knorr, N., Duran, R. S., Ingebrandt, S., Offenhäusser, A., Knoll, W., & Köper, I. (2005). Membrane on a chip: a functional tethered lipid bilayer membrane on silicon oxide surfaces. *Biophysical Journal, 89*, 1780–1788.
113. Wagner, M. L., & Tamm, L. K. (2000). Tethered polymer-supported planar lipid bilayers for reconstitution of integral membrane proteins: Silane-polyethyleneglycol-lipid as a cushion and covalent linker. *Biophysical Journal, 79*, 1400–1414.
114. Okada, T., & Ikada, Y. (1993). Tissue reactions to subcutaneosly implanted, surface modified silicones. *Journal of Biomedical Materials Research, 27*, 1509–1518.
115. Caracciolo, G., Palchetti, S., Colapicchioni, V., Digiacomo, L., Pozzi, D., Capriotti, A. L., La Barbera, G., & Laganà, A. (2015). Stealth effect of biomolecular corona on nanoparticle uptake by immune cells. *Langmuir, 31*, 10764–19773.
116. Wassef, N. M., & Alving, C. R. (1993). Complement-dependent phagocytosis of liposomes. *Chemistry and Physics of Lipids, 64*, 239–248.
117. Moghimi, S. M., Andersen, A. J., Hashemi, S. H., Lettiero, B., Ahmadvand, D., Hunter, A. C., Andresen, T. L., Hamad, I., & Szebeni, J. (2010). Complement activation cascade triggered by PEG-PL engineered nanomedicines and carbon nanotubes: the challenges ahead. *Journal of Controled Release, 146*, 175–181.
118. Hajipour, M. J., Safavi-Sohi, R., Shaifi, S., Mahmoud, N., Ashkarran, A. A., Voke, E., Serpooshan, V., Ramezankhani, M., Milani, A. S., Landry, M. P., & Mahmoudi, M. (2023). An overview of nanoparticle protein corona literature. *Small, 19*, e2301838.
119. Fernandez Oropeza, N., Zurek, N. A., Malvan-De La Cruz, M., Fabry-Wood, A., Fetzer, J. M., Graves, S. W., & Shreve, A. P. (2017). Multiplexed lipid bilayers on silica microspheres for analytical screening applications. *Analytical Chemistry, 89*, 6440–6447.
120. Chemburu, S., Fenton, K., Lopez, G. P., & Zeineldin, R. (2010). Biomimetic silica microspheres in biosensing. *Molecules, 15*, 1932–1957.
121. Li, B., Wang, F., Gui, L., He, Q., Yao, Y., & Chen, H. (2018). The potential of biomimetic nanoparticles for tumor-targeted drug delivery. *Nanomedicine (London, England), 13*, 2099–2118.
122. Oieni, J., Levy, L., Khait, N. L., Yosef, L., Schoen, B., Fliman, M., Shalom-Luxenburg, H., Dayan, N. M., D'Atri, D., Anavy, N. C., & Machluf, M. (2020). Nano-Ghosts: Biomimetic membranal vesicles, technology and characterization. *Methods, 177*, 126–134.
123. Meyer, R. A., Mathew, M. P., Ben-Akiva, E., Sunshine, J. C., Shmueli, R. B., Ren, Q., Yarema, K. J., & Green, J. J. (2018). Anisotropic biodegradable lipid coated particles for spatially dynamic protein presentation. *Acta Biomaterialia, 72*, 228–238.
124. Liu, L., Bai, X., Martikainen, M. V., Kårlund, A., Roponen, M., Xu, W., Hu, G., Tasciotti, E., & Lehto, V. P. (2021). Cell membrane coating integrity affects the internalization mechanism of biomimetic nanoparticles. *Nature Communications, 12*, 5726.
125. Gao, J., Dong, X., & Wang, X. (2020). Generation, purification and engineering of extracellular vesicles and their biomedical applications. *Methods, 177*, 114–125.
126. Luo, C., Hu, X., Peng, R., Huang, H., Liu, Q., & Tan, W. (2019). Biomimetic carriers based on giant membrane vesicles for targeted drug delivery and photodynamic/photothermal synergistic therapy. *ACS Applied Material Interfaces, 11*, 43811–43819.
127. Fernandes, L. F., Bruch, G. E., Massensini, A. R., & Frézard, F. (2018). Recent advances in the therapeutic and diagnostic use of liposomes and carbon nanomaterials in ischemic stroke. *Frontiers in Neuroscience, 12*, 453.

128. Wei, X., Ying, M., Dehaini, D., Su, Y., Kroll, A. V., Zhou, J., Gao, W., Fang, R. H., Chien, S., & Zhang, L. (2018). Nanoparticle functionalization with platelet membrane enables multifactored biological targeting and detection of atherosclerosis. *ACS Nano, 12*, 109–116.
129. Vasdev, N., Pawar, B., Gupta, T., Mhatre, M., & Tekade, R. K. (2023). A bird's eye view of various cell-based biomimetic nanomedicines for the treatment of arthritis. *Pharmaceutics, 15*, 1150.
130. Li, X., He, Y., Zhu, Y., Jiang, L., Zhang, S., Qin, J., Wu, Q., Dai, W., Shen, S., Pang, Z., & Wang, J. (2019). Route to rheumatoid arthritis by macophage-derived-microvesicle-coated nanoparticles. *Nano Letters, 19*, 124–134.
131. Chen, J., Zeng, S., Xue, Q., Hong, Y., Liu, L., Song, L., Fang, C., Zhang, H., Wang, B., Sedgwick, A. C., Zhang, P., Sessler, J. L., Liu, C., & Chen, J. (2022). Photoacoustic image-guided biomimetic nanoparticles targeting rheumatoid arthritis. *PNAS, 119*, e2213373119.
132. Lv, P., Liu, X., Chen, X., Liu, C., Zhang, Y., Chu, C., Wang, J., Wang, X., Chen, X., & Liu, G. (2019). Genetically engineered cell membrane nanovesicles for oncolytic adenovirus delivery: a versatile platform for cancer virotherapy. *Nano Letters, 19*, 2993–3001.
133. Wang, D., Liu, C., You, S., Zhang, K., Li, M., Cao, Y., Wang, C., Dong, H., & Zhang, X. (2020). Bacterial vesicle-cancer cell hybrid membrane-coated nanoparticles for tumor specific immune activation and photothermal therapy. *ACS Applied Material Interfaces, 12*, 41138–41147.
134. Lingling, W., Ding, H., Xin, Q., Shi, X., Yang, J., Huang, M., Zhang, J., Zhang, H., Song, J., Zhu, L., Song, Y., Ma, Y., & Yang, C. (2020). Fluidic multivalent membrane nanointerface enables synergetic enrichment of circulating tumor cells with high efficiency and viability. *Journal of the American Chemical Society, 142*, 4800–4806.
135. Jia, L., Zhen, X., Chen, L., Feng, Q., Yuan, W., Yusi, B., Wang, S., & Xie, X. (2023). Bioinspired nano-plate-coral platform enabled efficient detection of circulating tumor cells via the synergistic capture of multivalent aptamer and tumor cell membrane. *Journal of Colloid and Interface Science, 631*, 55–65.
136. Czaplicka, M., Nicincki, K., Nowicka, A., Szymborski, T., Chmielewska, I., Trzcinska-Danielewicz, J., Girstum, A., & Kaminska, A. (2020). Effect of varying expression of EpCAM on the efficiency of CTCs detection by SERS-based immunomagnetic optofluidic device. *Cancers, 12*, 3315.
137. Zhang, X., Jia, L., Liu, N., Zhao, Y., Zhang, T., & Xie, X. (2023). Inside-out extracellular vesicles-like biomimetic magnetic nanoparticles for efficient screening P-Glycoprotein inhibitors to overcome cancer multidrug resistance. *Colloids Surfaces B Biointerfaces., 222*, 113134.
138. Navarro-Palomares, E., García-Hevia, L., Padín-González, E., Bañobre-López, M., Villegas, J. C., Valiente, R., & Fanarraga, M. L. (2021). Targeting nanomaterials to head and neck cancer cells using a fragment of the shiga toxin as a potent natural ligand. *Cancers (Basel), 13*(19), 4920.
139. Liszewski, M. K., & Atkinson, J. P. (2023). Alternative pathway activation: Ever ancient and ever new. *Immunological Reviews, 313*, 60–63.
140. Ye, J. J., Bian, X., Lim, J., & Medzhitov, R. (2020). Adiponectin and related C1q/TNF-related proteins bind selectively to anionic phospholipids and sphingolipids. *PNAS, 117*, 17381–17388.
141. Kim, H. H., Kim, K., & Byun, Y. (2005). Preparation of a chemically anchored phospholipid monolayer on an acrylated polymer substrate. *Biomaterials, 26*, 3435–3444.
142. Kochanowski, A., Hoene, A., Patrzyk, M., Walschus, U., Finke, B., Luthringer, B., Feyerabend, F., Willumeit, R., Lucke, S., & Schlosser, M. (2011). Examination of the inflammatory response following implantation of titanium plates coated with phospholipids in rats. *Journal of Materials Science Materials in Medicine, 22*, 1015–1026.
143. Her, C., Filoti, D. I., McLean, M. A., Sligar, S. G., Ross, J. B. A., Steele, H., & Laue, T. M. (2016). The charge properties of phospholipid nanodiscs. *Biophysical Journal, 111*, 989–998.

144. Ishihara, K., Ueda, T., & Nakabayashi, N. (1990). Preparation of phospholipid polymers and their properties as polymer hydrogel membranes. *Polymer Journal, 22*(5), 355–360.
145. Iwasaki, Y., & Ishihara, K. (2012). Cell membrane-inspired phospholipid polymers for developing medical devices with excellent biointerfaces. *Science and Technology of Advanced Materials, 13*(6), 064101.
146. Amoako, K., Ukita, R., & Cook, K. E. (2025). Antifouling zwitterionic polymer coatings for blood-bearing medical devices. *Langmuir, 41*, 2994–3006.
147. Li, P., Li, X., Cai, W., Chen, H., Chen, H., Wang, R., Zhao, Y., Wang, J., & Huang, N. (2020). Phospholipid-based multifunctional coating via layer-by-layer self-assembly for biomedical applications. *Materials Science and Engingeering C, 116*, 111237.
148. Hou, A., He, L., Han, M., Shi, S., Cheng, L., Luo, J., Jia, L., Jiy, L., & Yang, J. (2024). Engineering coatings inpired by cell membrane thermal dynamics to enhance photothermal therapy and osteogenesis for implant infections. *ACS Applied Materials Interfaces, 16*, 53425–53448.

Chapter 5
Back to the Future: Perspectives for Biomimetic Membrane in Nanomedicine

The purpose of this book is to draw biomimetic membrane technology into one of the central themes of nanobiotechnology for the development of novel medical diagnostic and therapeutic devices and hence to elucidate the significant role that biomimetic membrane technology has for the growth of the field of nanomedicine. There is a dichotomy of whether the starting point for advances in nanobiotechnology and nanomedicine lies in the realm of fundamental scientists to first seek knowledge in the basic functions of molecules that comprise biomimetic membranes or in the realm of clinicians and engineers who first seek to observe and characterize defined medical problems and needs. In those scenarios, the fundamental scientist will have the subsequent task to find a biomedical application of such fundamental knowledge and understanding. The clinician or engineer will have the subsequent task to find, or indeed reverse-engineer, a targeted understanding of the functions and principles that underlie the defined problem. This question of the preeminence of fundamental or applied research will continue to be the source of debate amongst scientists, clinicians, and engineers.

This chapter proposes to provide a capstone to this book by taking a "back-to-the-future" perspective, where we consider that patients also have a major role to play in nanomedicine. This role is in accepting and trusting that a medical device, either diagnostic or therapeutic, will function efficaciously with their body. This is a major role in the paradigm shift in nanomedicine we introduced in Chap. 3 for the development of "symbio-bots" to function symbiotically and autonomously with the body of the patient. Is such a paradigm shift really only a restatement of common sense? Most likely the answer to this question is "yes." Nonetheless, the detail required to develop functional "symbio-bots" does need a major advance in the science and engineering of biomimetic membrane technology to improve the interface of the "symbio-bot" with the body.

D. Martin, I. Vilgrain, *Biomimetic Membrane Technology*,
https://doi.org/10.1007/978-3-031-90594-0_5

Back to the Future or Return to the Past?

The concept of a symbiotic medical device could arguably be traced to ancient Egyptian medical practices, where evidence has been discovered of an intravital amputation of a big toe and replacement by a wooden prosthesis [1]. This intravital amputation and prosthesis were discovered during the examination of a mummy archeologically dated to the 21st/22nd dynasty (*ca.* 1065–740 BC) excavated from a tomb at Thebes-West (Sheik-Abd-el-Gurna). The mummy was of a 50–55-year-old woman whose big toe of the right foot had been amputated during her lifetime. The amputated toe had been replaced by a wooden prosthesis, painted dark brown and made up of three separate components (Fig. 5.1). The first component was a large wooden corpus carved into the shape of the big toe. The second component comprised two wooden plates. The third component was a textile lace that fixed the replacement toe to the wooden plates and then firmly to the foot [1]. Apart from the aesthetic considerations of having a prosthetic big toe, there is significant functional assistance from such a prosthetic big toe since it usually bears around 40% of the walking weight and hence would have been important for the woman's physical integrity.

Over the ages since that ancient Egyptian big toe, several types of limb and appendage prostheses have been utilized to replace lost form. There are numerous examples of such prostheses, but here we illustrate an example from the seventeenth century (Fig. 5.2).

Nonetheless, interfacing such prostheses to the nerve, muscle, and cardiovascular system provides major challenges to be overcome in order to replace lost function. The intention to create a symbiotic prosthesis to replace lost form is a principle that could be considered as a return to the past. However, building upon that principle brings us back to the future, where we utilize the advances in nanobiotechnology, biomimetic membrane technology, and biological engineering to develop the interface of a symbiotic prosthesis to create a "symbio-bot" that replaces, or indeed enhances, lost function. The development of a "symbio-bot" benefits from the role played by the patient to accept that the "symbio-bot" will function symbiotically and autonomously with the body.

A Roadmap for Biomimetic Membrane Technology in Nanomedicine

We can conclude by summarizing the remarks made at the end of Chap. 3 by returning to the context of how the role of the patient and the concepts of human factors can be integrated with the technical development of a "symbio-bot." We have introduced the need for the Societal Readiness Level (SRL) to be evaluated concurrently with the Technology Readiness Level (TRL). The significant opportunity in framing this new health paradigm for "symbio-bots" is to challenge all the scientific

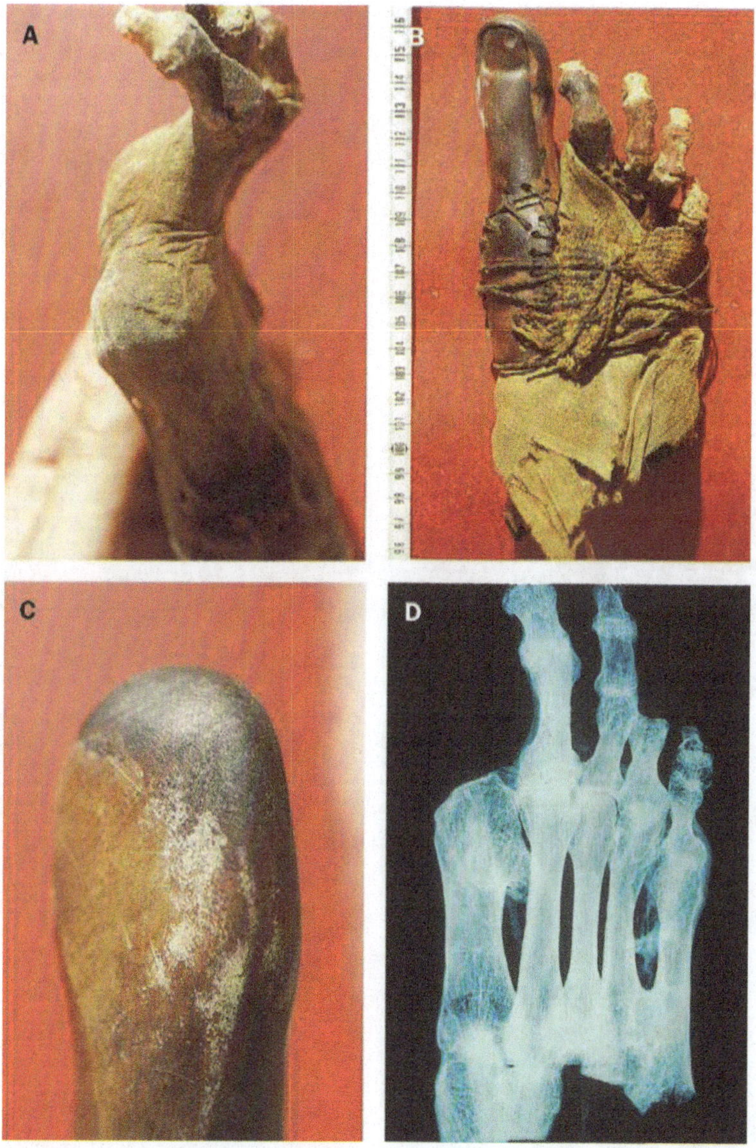

Fig. 5.1 Views of right foot and prosthesis, reproduced with permission from Ref. [1]. (**a**) well-healed amputation area covered by intact layer of skin (**b**) wellcrafted wooden prosthesis, attached to the forefoot by a textile lace (**c**) basal surface of prosthesis showing abrasion of the wooden surface and several scratch marks indicating intravital use (**d**) radiograph showing bony overgrowth of the first metatarsal indicating longstanding intravital amputation of the big toe. Note the enhanced demineralisation of the first metatarsal

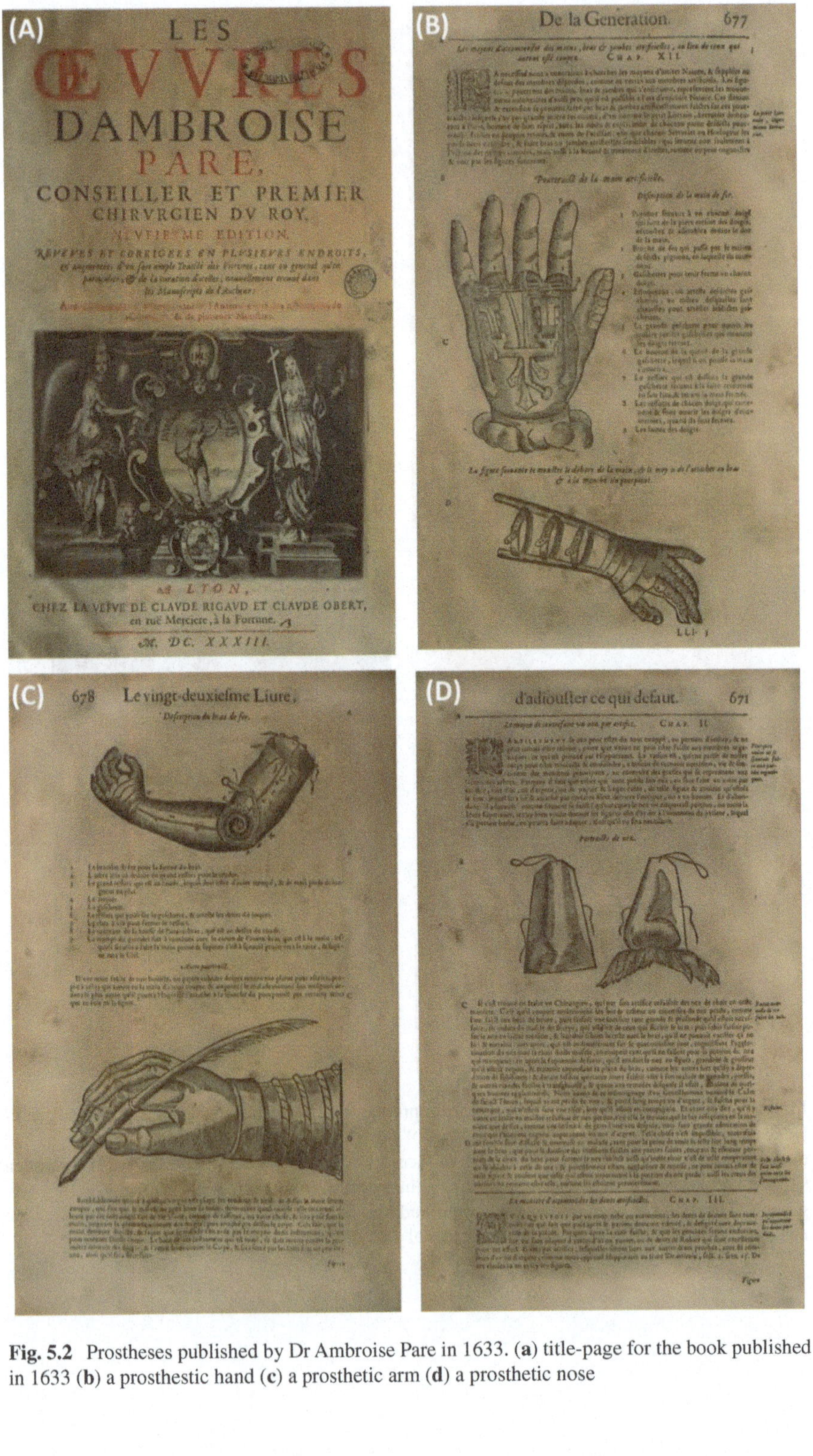

Fig. 5.2 Prostheses published by Dr Ambroise Pare in 1633. (**a**) title-page for the book published in 1633 (**b**) a prosthetic hand (**c**) a prosthetic arm (**d**) a prosthetic nose

disciplines of humanities and social sciences (SHS). This is particularly important in constructing the philosophical and ethical framework for developing expertise in UX design, including co-design with patients, and in corporal/emotional communication and design. The explicit response for "symbio-bot" is the process of understanding and appropriating this new paradigm for patients, to then demonstrate the relevance of multidisciplinarity at theoretical and methodological complementarity. A challenge is to demonstrate whether technologies like "symbio-bots" transform the habits and expectations of patients, which have been traditionally built around the natural limits of the human being and body that are intuitively obvious to people. Such limits are also shifted so that a new paradigm of symbiotic medical devices, such as "symbio-bots," takes on a meaning with regard to a realization of the model of the vision of medicine as predictive, preventive, personalized and participatory ("P4 medicine").

Those are the same advantages for biological engineered smart nanostructured systems designed to function as implanted medical devices. For an implanted application, the additional, and highly significant, advantage of utilizing biological components is the ease of providing a more natural interface to living tissues. This is because a smart nanostructured system based on biological engineering utilizes biological components that include lipids and proteins. We return to the concept of the "symbio-bot" introduced in the previous section. The "symbio-bot" embodies the advantages of utilizing a biological engineering approach to develop an implanted medical device.

The major challenges to overcome for the function of an implanted "symbio-bot" are the biocompatibility and the interface of the implanted device with the body. In Chap. 2, we outlined a roadmap for the design of an implantable "symbio-bot" based on biological engineering. This type of roadmap would be useful, for example, to increase the longevity of bioelectrodes that are designed for implantation in the body. It is clear that a balance needs to be struck with the *"technology or materials push"* and the *"medical device pull"* that is based on clinical needs and questions. Nonetheless, such a *"medical device pull"* will need the skills of materials and engineering disciplines to design the "symbio-bot" that is required to solve the clinical needs. The major questions to be answered at the various stages of this roadmap to *"pull"* the medical device are:

1. What is the clinical problem to be solved?
2. Is the diagnostic device functioning inside the body or using fluid or tissue samples from the body?
3. Is the diagnostic device easily sterilized and does it pass regulatory affairs?
4. Does the diagnostic device actually help clinicians with medical diagnosis?
5. Is a diagnostic device needed, or is pharmaceutical therapy preferred?
6. Do patients understand why the diagnostic device is needed, and is it tolerated by patients?

The progression along this roadmap requires at least the interdisciplinary expertise of clinicians, engineers, biologists, biochemists, biophysicists, materials scientists, and regulatory affairs professionals. To follow this roadmap from patient to device or pharmaceutical treatment requires a multidisciplinary approach, where the primary driver is the needs of the patient.

The vision outlined for the concept of the "symbio-bot" is predicated on two questions, which are at the basis for transforming the paradigm for medical devices and health. The first question is at the level of the patient, who needs to understand, consider, and accept a specific medical symbiotic device that is intended to be used inside their body. The second is at the level of the medical device, which needs appropriate technology so that it understands, considers, interacts with, and is accepted by the body tissues of the patient.

Reference

1. Nerlich, A. G., Zink, A., Szeimies, U., & Hagedorn, H. G. (2000). Ancient Egyptian prosthesis of the big toe. *Lancet, 356*, 2176–2179.

Index

D. Martin, I. Vilgrain, *Biomimetic Membrane Technology*,
https://doi.org/10.1007/978-3-031-90594-0

The manufacturer's authorised representative in the EU is Springer
Nature Customer Service Centre GmbH, Europaplatz 3, 69115 Heidelberg,
Germany. If you have any concerns regarding our products, please
contact ProductSafety@springernature.com

Printed and bound by CPI Group (UK) Ltd, Croydon, CR0 4YY
30/09/2026
02225259-0011